RENEWALS 458-4574
DATE DUE

**WITHDRAWN
UTSA LIBRARIES**

GROUNDWATER FLOW UNDERSTANDING
FROM LOCAL TO REGIONAL SCALE

SELECTED PAPERS ON HYDROGEOLOGY

12

Series Editor: Dr Nick S. Robins
Editor-in-Chief IAH Book Series
British Geological Survey
Wallingford, UK

 INTERNATIONAL ASSOCIATION OF HYDROGEOLOGISTS

Groundwater flow understanding
from local to regional scale

Edited by

J. Joel Carrillo R.
Institute of Geography, UNAM, Coyoacán, México

M. Adrian Ortega G.
Centre for Geosciences, UNAM, Querétaro, México

Taylor & Francis
Taylor & Francis Group

LONDON / LEIDEN / NEW YORK / PHILADELPHIA / SINGAPORE

Taylor & Francis is an imprint of the Taylor & Francis Group, an informa business

© 2008 Taylor & Francis Group, London, UK

Typeset by Charon Tec Ltd (A Macmillan Company), Chennai, India
Printed and bound in Great Britain by Antony Rowe Ltd (A CPI-group Company), Chippenham, Wiltshire

All rights reserved. No part of this publication or the information contained herein may be reproduced, stored in a retrieval system, or transmitted in any form or by any means, electronic, mechanical, by photocopying, recording or otherwise, without written prior permission from the publishers.

Although all care is taken to ensure integrity and the quality of this publication and the information herein, no responsibility is assumed by the publishers nor the author for any damage to the property or persons as a result of operation or use of this publication and/or the information contained herein.

Published by: Taylor & Francis/Balkema
               P.O. Box 447, 2300 AK Leiden, The Netherlands
               e-mail: Pub.NL@tandf.co.uk
               www.balkema.nl, www.taylorandfrancis.co.uk, www.crcpress.com

British Library Cataloguing in Publication Data
A catalogue record for this book is available from the British Library

Library of Congress Cataloging in Publication Data

Groundwater flow understanding: from local to regional scale/edited by J. Joel Carrillo R., M. Adrian Ortega G.
   p. cm. — (Selected papers on hydrogeology/International Association of Hydrogeologists; 12)
   Includes bibliographical references and index.
   ISBN 978-0-415-43678-6 (hardcover: alk. paper)
   1. Groundwater flow. 2. Hydrogeology. I. Carrillo-Rivera, J. J. II. Ortega G., M. Adrian (Ortega Guerrero)

GB1197.7.G7646 2008
551.49—dc22

                                                             2007024300

ISBN-13 978-0-415-43678-6

ISBN-13 (e-book): 978-0-203-94579-7

Contents

Preface vii
About the editors xi

Keynote Lecture

Evolution of palaeowaters in sedimentary basins and coastal aquifers;
valuable natural resources and archives of climatic and environmental change 1
Mike W. Edmunds

Groundwater flow system, subsidence and solute transport controls in the
lacustrine aquitard of Mexico City 9
M.A. Ortega-Guerrero

Groundwater flow system response in thick aquifer units: theory and practice in Mexico 25
J.J. Carrillo-Rivera and A. Cardona

Chapter 1. Integrative modelling of global change effects on the water cycle
in the upper Danube catchment (Germany) – the groundwater
management perspective 47
Roland Barthel, Wolfram Mauser and Juergen Braun

Chapter 2. Water management in transboundary hard rock regions – A case study
from the German-Czech border region 73
S. Bender, T. Mieseler, T. Rubbert and S. Wohnlich

Chapter 3. Combined use of indicators to evaluate waste-water contamination
to local flow systems in semi-arid regions: San Luis Potosi, Mexico 85
*A. Cardona, J.J. Carrillo-Rivera, G.J. Castro-Larragoitia and
E.H. Graniel-Castro*

Chapter 4. Integrating physical hydrogeology, hydrochemistry, and environmental
isotopes to constrain regional groundwater flow: southern
Riverine Province, Murray Basin, Australia 105
Ian Cartwright, Tamie R. Weaver and Sarah O. Tweed

Chapter 5. The South African groundwater decision tool 135
Ingrid Dennis and Sonia Veltman

VI *Contents*

Chapter 6. Causes and implications of the drying of Red Rock crater lakes, Australia   151
R. Adler and C.R. Lawrence

Chapter 7. The development of a methodology for groundwater management
in dolomitic terrains of South Africa   165
S. Veltman and B.H. Usher

Author index   183

Subject index   185

SERIES IAH-Selected Papers   187

Preface

By J. Joel Carrillo R. and M. Adrian Ortega G., Invited Editors

Any sustainable groundwater development programme requires an understanding of the flow system. This a prerequisite to understanding both groundwater availability and the dependence between groundwater and other components of the environment. This awareness can be achieved through groundwater flow understanding: from local to regional scales. The understanding of groundwater flow in its relevant scale is essential for studies involving engineering, geography, agriculture, ecology, and in a broad sense, any environmental related issue.

The impetus for this book came from the XXXIII International Hydrogeologic Congress organized by the IAH-Mexican Chapter in Zacatecas, Mexico in 2004. This Congress focused on Groundwater Flow Understanding from Local to Regional Scales. The response to different sessions in the Congress, both in the number of abstracts received and in attendance, indicated that a wide range of research activity worldwide is focused on different scale research and applications, where the understanding of the groundwater flow systems plays an important role.

The congress was aimed at groundwater researchers and professionals, students, water resources specialists, government administrators and educators, and those interested in groundwater and the environment. It was a forum for exchanging techniques, knowledge, ideas and experience with groundwater studies and investigations. Another aim of the congress was to gain a better communication with the general public and non-groundwater specialists.

The objectives of the Congress were:

- Exchange experiences on integral groundwater management with scale flow issues.
- Propose methods of defining, preventing, controlling and mitigating negative environmental impacts related to groundwater.
- Discuss specific issues such as trans-boundary groundwater flow, groundwater recharge, groundwater mining, groundwater flow in thick aquifers.
- Communicate effectively with the general public and non-groundwater specialists.
- Consider the importance of sustainable development of groundwater, and its social and economic implications.
- Present recommendations to administrators and professionals responsible for water management.

There were nine main topics in the Congress:

1. **Environmental issues of groundwater-flow scaling.** Management of groundwater or of the environment, may produce an impact locally, where it was applied, or at a distance. Control of these impacts requires understanding of the interaction between flow systems, the environmental response and its extension (local or regional). Papers were presented on groundwater flow system scaling under different hydrogeological scenarios, extensive groundwater exploitation and environmental impacts such as: deterioration of ecosystems, land subsidence and ground surface fracturing, and progressive impairment of groundwater quality. Other papers described the impacts and land use processes or

productive activities (including agricultural, animal waste, mining, oil production, industry, urban) on groundwater contamination at different scales of application.

2. **Chemical and isotopic data in local and regional groundwater flow definition.** Physical, chemical and isotopic characteristics of groundwater are paramount to understanding groundwater flow. Papers were presented in which major ions, trace elements, tracer tests, stable and radioactive isotopes are used to determine origin, residence time and chemical evolution of groundwater, definition of recharge zones and estimation of storage volumes, and the general hierarchy or extension of groundwater flow (local, intermediate, regional). Other papers described fresh water-saline water interaction in both inland and coastal aquifers; anthropogenic impacts on groundwater sources; hydrochemistry of thermal systems; and hydrochemistry and associated cycles of critical contaminants for human health.

3. **Groundwater flow scaling in hard-rock media.** Contributions included descriptions of hard-rock units, including mainly basement rock or weathered basement intrusive and metamorphic rocks. Studies on extrusive and sedimentary rock units, which act in a similar manner, were also presented. Application of geophysical methods to fracture characterization; assessment of groundwater flow and contaminant transport were the focus of other presentations in hard rock media.

4. **Role of flow systems in contaminant migration.** Issues include flow direction and movement velocity of a contaminant in the subsoil, site characterization and remedial approaches, as well as natural or artificial attenuation. The factors influencing the movement and fate of contaminants and the physical, bacteriological and chemical characteristics of the geological media were analyzed under different scale conditions of groundwater flow. Different contaminant sources (urban, industrial and agrochemical) and field tracer experiments were also assessed. Behaviour and transformation of inorganic, organic and biological contaminants were analyzed. Many of the papers include aspects of groundwater resources protection.

5. **Recharge to local and regional systems.** Papers were presented on the processes to enhance groundwater sources through artificial recharge and induced recharge by increasing the efficiency of natural recharge processes. Hydrogeologic investigations to determine the technical feasibility for recharge at selected sites, response of artificial recharge and induced recharge evaluation and identification of natural recharge were discussed. Particular attention was given to recharge mechanisms and the importance of the associated flow system, changes in chemistry or physical characteristics of water or aquifer matrix as well as the identification of the importance of the geological framework.

6. **Wetlands and groundwater flow dimensions.** The survival of groundwater dependent wetlands and related ecosystems depends on continuous groundwater inflow; therefore, clear understanding of the source of water and its flow path is required to ensure preservation of the wetland. Papers on the identification, understanding and management of groundwater inflow to wetlands according to water quality, hydrodynamics and the groundwater flow path were presented.

7. **Differential groundwater flow to coastal areas.** Groundwater inflow from the continent to coastal areas is related to the amount of freshwater that controls the lateral movement of seawater intrusion. Papers focused on groundwater inflow equilibrium between the ecosystems in coastal areas, based on adequate management of natural resources inland.

8. **Modelling of groundwater flow systems.** Modern numeric modelling techniques allow the inclusion of hydrological and hydrogeological properties obtained by both direct and indirect methods to produce improved simulations of groundwater conditions. Papers showed research results regarding the various flow system components by means of numeric modelling and its multidisciplinary applications.
9. **Flow systems: social, legal, economical and educational aspects of groundwater management.** Papers showed the interest of teachers, managers, legislators, economists and the general public in understanding groundwater behaviour. Different examples of educational, legislative as well as participative aspects at all levels were based on an understanding of the flow systems.

In this Congress, professionals from all over the world met to exchange experiences and methods, to discuss problems and their solutions, and to review the application of techniques that involve the perspective of scale in groundwater movement.

The papers contained in this volume are representative of the research currently being conducted in environmental applications on local and regional scales of groundwater flow system. The papers represent an excellent cross section of critical environmental issues related to groundwater flow systems in terms of their physical, chemical and biological interaction. This book raises the following questions:

1. Is there an adequate understanding of suitable information regarding how a hydrologic budget represents the functioning of a surface basin?
2. Where and how does recharge and discharge take place?
3. What is the influence of a complex geological framework in the flow and chemistry of groundwater?
4. Which part of the flow system is highly susceptible to contamination, or conversely is a favoured option for waste disposal?
5. Which areas of the flow system require special management and protection to preserve long-term drinking water quality conditions?
6. How are other components of the environment affected by changes in the groundwater flow system?

About the editors

Adrian Ortega is Professor at the National University of Mexico. His research interest focuses on groundwater flow, origin, chemistry and solute and contaminant transport in high compressible closed basin aquitards, from detailed local field instrumentation sites to watershed scale.

J. Joel Carrillo R. is Professor of groundwater at the Institute of Geography, UNAM (México) and obtained relevant experience in Australia. He and his students research a variety of hydrogeological issues related to the groundwater flow system analysis applied to the definition of environmental problems. He is the President of the IAH-Mexican Chapter.

KEYNOTE LECTURE

Evolution of palaeowaters in sedimentary basins and coastal aquifers; valuable natural resources and archives of climatic and environmental change

Mike W. Edmunds
Oxford Centre for Water Research, School of Geography and Environment, Mansfield Road, Oxford OX1 3TB

ABSTRACT: Groundwater contained particularly in large sedimentary basins contains an important overall record of hydrogeological evolution, and forms a direct archive of past climatic and hydrological change in the late Pleistocene and Holocene. Combinations of a range of isotopic and chemical fingerprints may be used to follow sequential changes along flow pathways in large basins. The large sedimentary basins of northern Africa, described here, illustrate the dramatic changes that took place through the last glacial maximum to the warming and impacts at the continental scale of global sea level rise in the early Holocene. Complimentary changes are found in the coastal and offshore areas of Europe where lowered sea levels until some 8,000 yr BP allowed development of groundwater systems beyond modern coastlines.

Keywords: Palaeowater, formation water, climate change, isotopes, hydrochemistry, coastline.

1 INTRODUCTION

The evolution of fresh groundwater is a process that takes place continuously from the time of sedimentary deposition, or lithification of hard rock igneous bodies, right up to the present day. Connate waters (*sensu stricto*) of marine or continental origin are progressively modified by successive cycles of groundwater flow, which are controlled by various forces, among them tectonic movements and changing crustal stress patterns, changing geothermal gradients, eustatic changes, glacial stress and shifting climate patterns which control recharge and the flux to groundwater. The associated freshwater diagenesis of the evolving sedimentary or igneous aquifer materials is an ongoing process and may be followed at the present day through signatures in both the water and in the rock. Exploited fresh groundwater is more than likely to contain residues of connate water, modified formation water or more recent '*palaeowater*', in addition to active recharge forming the most recent contribution to the present-day hydrological cycle. These processes of evolution may be recognised through the fingerprints of the inert and reactive tracers, which make up the overall groundwater quality.

Groundwater contained in large sedimentary basins contains an important overall record of hydrogeological evolution, but is emerging in particular as a *direct* archive of past climatic and hydrological change in the late Pleistocene and Holocene, which may be used alongside other proxy data. Indirect evidence of the palaeohydrology at both low and mid-latitude, has been deduced from various sources especially the geomorphological record, dated lake sediments (Gasse 2000) and speleothems. In contrast to other archives such as ice cores or tree rings, which contain high-resolution information, data available in large groundwater bodies are of low resolution (typically ± 1,000 yr). This is due to the advection or dispersion of input signals in the water body. Additionally many groundwater data are obtained from pumped samples where sample intervals may extend over tens of metres and records are destroyed; the preservation of a groundwater 'stratigraphy' has been widely demonstrated. This paper highlights recent and not so recent studies that show the importance of groundwater as an archive using examples from North Africa and coastal Europe.

2 RECOGNITION OF PALAEOWATERS

Palaeowater (a term coined during the first studies of dated Saharan groundwater in the early 1960's) can be defined by isotopic or other criteria as having evolved mainly during cooler climatic conditions of the late Pleistocene (Edmunds et al., 2001). They can be either relatively immobile bodies, or part of the main flowing groundwater systems (figure 1). The largest continental sedimentary aquifer system such as the Great Artesian Basin in Australia have turnover times in excess of 400 kyr (Lehmann et al., 2003), determined using the long lived noble gas tracer ^{81}Kr, and may therefore contain palaeowater relating to successive Pleistocene climatic cycles. More recently residence time up to 1 M year has been reported from the Western Desert of Egypt using ^{81}Kr and ^{36}Cl (Sturchio et al., 2004) Specific climate-related information may be obtained from the isotopic and

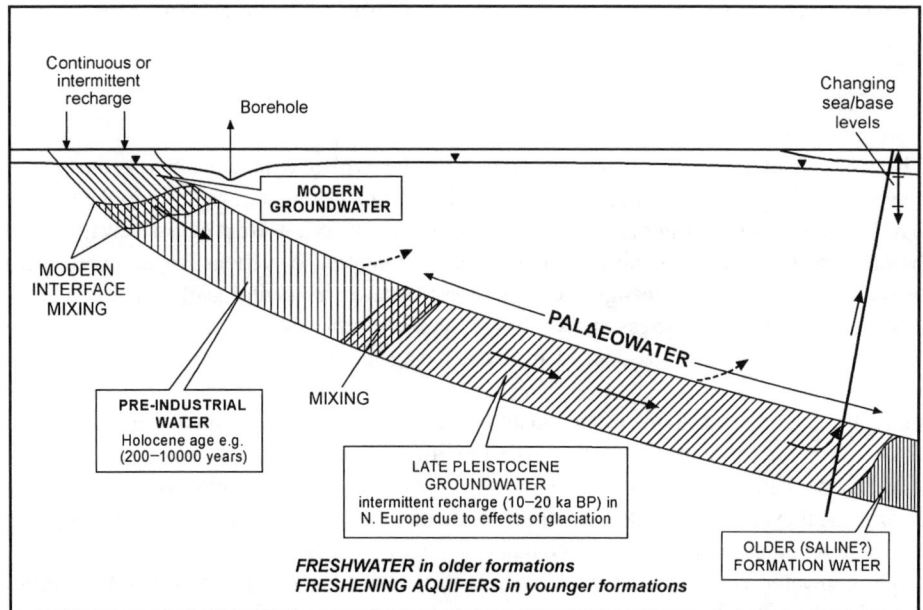

Figure 1. Conceptual model to show the typical relationships between modern waters, palaeowaters and saline formation waters.

chemical records in groundwater including palaeotemperature, past precipitation amount as well as evapotranspiration, patterns of former air mass circulation and continentality. The stable isotope records ($\delta^{18}O$, $\delta^{2}H$) in modern rainfall are now well understood at a global scale and the basis exists for interpretation of precipitation characteristics of past climates (Rosanski et al., 1997). The ^{36}Cl content of groundwater may also be used in low chloride water to deduce the former composition of atmospheric deposition as well as Late Pleistocene evapotranspiration (Andrews et al., 1994). Noble gas ratios also have now been well established as reliable tools for measuring past groundwater temperature (Loosli et al., 1999). Radiocarbon, although a reactive tracer, remains the principle tool for groundwater residence time studies. Corrected ages however may be difficult to obtain due to the need to define the reactants involved and to fully understand the carbon hydrogeochemistry and in this paper values as percent modern carbon (pmc) are used.

Both inert and reactive geochemical tracers, as well as isotope signatures, may provide evidence of past environmental and climate change. The range in chloride concentrations may be equated with past and present day recharge/evaporation cycles in many environments. Many large sedimentary basins have remained aerobic for 20 kyr or more and under these conditions nitrate remains stable and may indicate evidence of past vegetation (Edmunds et al., 2004). Sequential water-rock interactions taking place as water moves down-gradient produce time-dependent records in their changing solute chemistry and these hydrogeochemical changes may in favourable circumstances also be used as qualitative residence time indicators and support absolute chronometers.

3 PALAEOWATER IN NORTH AFRICA

Dated groundwater from northern Africa illustrates the timing and evolution of aquifer recharge in the period since 30 kyr to the present (figure 2). Major climatic reorganisations controlled by shifts in the Atlantic westerly flows and the African monsoon can be recognised in groundwater of late Pleistocene and Holocene age; the Last Glacial Maximum (LGM) is recognised in north east Africa and the present day Sahel as a period of aridity with an absence of recharge, although recharge continued near to the Atlantic coast (Edmunds et al., 2004).

Examples are shown of the isotopic record from West Africa (Senegal, Mali and Morocco) as compared with north-east Africa (Libya and Egypt). There is a uniformity of composition of around $-6‰$ $\delta^{18}O$ in the groundwater from West Africa from the late Pleistocene to the present and the record shows a continuity of recharge over the period. This illustrates the relatively constant moisture source from the Atlantic. However if the coastal sites are examined, notably in Morocco it can be seen that the Holocene isotopic composition is lighter (more negative) than in the late Pleistocene. This small depletion in the $\delta^{18}O$ is related to the change in ocean volume and the release of large quantities of water from the continental ice sheets. A similar effect is also observed in coastal Portugal (Condesso de Melo et al., 2001).

The uniformity of isotopic composition and recharge contrasts with that in north east Africa where all waters are isotopically light as compared with groundwater nearer the Atlantic moisture source. Groundwater in the Kufra Basin (Nubian Sandstone) is isotopically the lightest in north Africa ($-11.5‰$ $\delta^{18}O$) and this contrasts with the Sirte Basin to the north where the palaeowater is some 3‰ more enriched.

Thus each sedimentary basin seems to have a distinctive composition, which supports the concept of local evolution of groundwater. Groundwater from the Egyptian oases lies

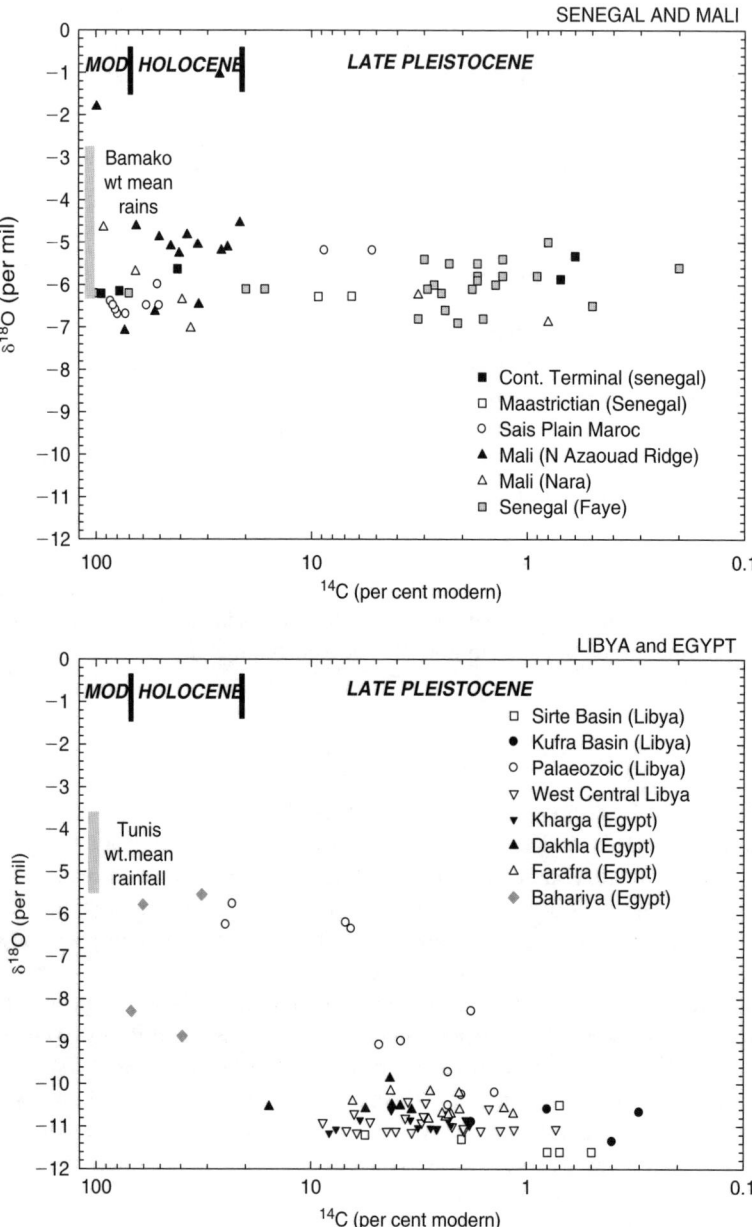

Figure 2. Isotopic compositions of dated groundwater from western and eastern regions of North Africa.

within the range −10 to −11‰ $\delta^{18}O$, distinct from the Sirte Basin at the same latitude and more akin to Kufra Basin composition. Although the continental effect could have led to the easternmost enrichments as seen in Egypt, the vast reserves of fresh groundwater are also anomalous in that they are found at the extreme of the evolution of the Atlantic air mass source where lower rainfall amounts would be expected. An additional possibility is that some recharge from the south-west within the hydraulically continuous Nubian sandstone took place, with a superimposed altitude effect the result of heavier rains and surface runoff from the Tibesti Mountains. A distinct arid interlude is also indicated by the absence of dated groundwater at the end of the late Pleistocene, corresponding to the period of aridity associated with the LGM.

The rapid global warming and sea level rise in the early Holocene led to intense periodic (ca 1,000 yr) wet cycles. This is recognisable in the groundwater record as a renewal of diffuse recharge and local river-induced recharge. These effects are particularly distinct in groundwater from Sudan where the signature of the intensification of the African monsoon can be seen. With the onset of the modern essentially arid climate of the present day, which became established around 4,500 yr BP, little or no recharge has taken place and aquifer systems, currently discharging in oases and sebkhats, are declining hydrogeological systems. The clear palaeowater signatures they contain present obvious signs of caution for groundwater management.

Chloride concentrations in most semi-arid groundwater may be equated with past and present day recharge/evaporation cycles and therefore chloride must be regarded as an independent palaeoenviromental indicator. The very fresh composition of continental formation water indicates evolution from long-duration wet periods. Moreover, many large sedimentary basins have remained aerobic for 20 kyr or more and under these conditions nitrate remains stable. The large concentration of nitrate, often exceeding potable guidelines, are taken to indicate evidence of past leguminous vegetation cover (Edmunds and Gaye, 1997).

4 PALAEOWATERS IN COASTAL AQUIFERS

Near modern coastlines bounded by sedimentary basins fresh palaeowater is now being recorded, trapped by the rapid Holocene sea level rise (around 8.5 kyr BP). Brackish or fresh waters may also be preserved offshore (Edmunds et al., 2001). Sea levels were lowered globally for up to 100 kyr during the Devensian period, allowing prolonged (re)activation of evolving groundwater flow systems, flushing of connate water and permitting the emplacement of freshwater at depth beneath and beyond present coastline as well as inland. This is well illustrated by several aquifers in Europe as well as the eastern seaboard of the USA and must also be a global phenomenon. Drilling of Pleistocene and Miocene sediments on the Atlantic coastal plain of the USA has proved the existence of fresh water ($<5,000$ mgL^{-1}) to depths of -200 m OD and as far as 100 km offshore (Hathaway et al., 1979). From glacial and sedimentary records it is now clear that many glacial episodes have affected the northern hemisphere during the past 1.7 Ma. Related fresh groundwater recharge events are therefore likely to have been cyclic over the recent geological past.

In Estonia the Cambrian-Vendian sandstone aquifer, which outcrops in the Gulf of Finland, contains freshwater to a depth of 300 m. This groundwater contains an extremely light isotopic composition (down to -22 ‰ δ^{18}O). This composition, one of the lightest, if not the lightest groundwater isotope value recorded, must originate from the Baltic ice sheet (Vaikmae et al., 2001). This was probably achieved by direct recharge under hydrostatic loading of melt-waters beneath the ice sheet perhaps aided by the buried glacial sedimentary channels incising the confined aquifer.

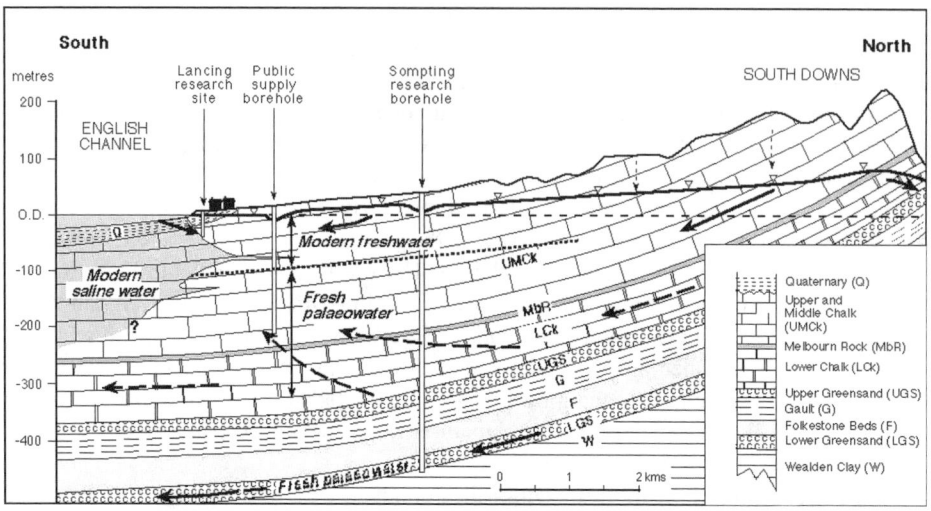

Figure 3. Cross section of the Channel coast of England near Brighton to show the detail of the coastal salinity and fresh palaeowaters beneath the present day groundwater circulation.

In the UK hydrogeophysical logging and geochemical sampling have identified fresh palaeowater to depths of 300 m beneath the Chalk aquifer, which outcrops on the south coast of England on the English Channel (figure 3). However, active groundwater circulation as shown by the flow and temperature profiles is restricted to the topmost 120 m of the aquifer. Evidence is found that fresh water in both the Cretaceous Chalk and the Albian sandstone aquifers extends offshore (Edmunds et al., 2001).

The coastal Aveiro Cretaceous aquifer, Portugal (Condesso de Melo et al., 2001) is probably exposed on the steeply dipping continental shelf allowing groundwater discharge especially during times of lowered sea level. Groundwater isotopic signatures and chemistry confirm this, containing a smooth radiocarbon increase and a flow sequence covering the late Pleistocene and Holocene. Noble gas ratios indicate that cooling at the time of the LGM was also around 6°C, as in northern Europe and in Africa. The stable isotope ratios on the other hand indicate an enrichment of 0.8–1.0 ‰ in $\delta^{18}O$ (as in north-west Africa but in contrast to the isotopic depletion found in northern Europe). This is interpreted to indicate that there has been constant air mass circulation at this latitude, but that the small difference reflects the composition changes in the oceans (in response to the ice volume changes).

5 CONCLUSIONS

The results discussed here make clear the importance for groundwater management to have a good understanding of groundwater residence times, their flow and recharge history, overall hydrogeology, as well as associated time-dependent geochemical changes. Palaeowater by definition is demonstrably free of human contaminants and must be regarded as high value groundwater which must be protected.

REFERENCES

Andrews, JN; Edmunds, WM; Smedley, PL; Fontes, J-Ch; Fifield, LK and Allan, GL (1994) Chlorine-36 in groundwater as a palaeoclimatic indicator: the East Midlands Triassic aquifer (UK). Earth and Planetary Science Letters, 122:159–172.

Condesso ed Melo, MT; Carreira-Paquette, PMM and Marques de Silva, MA (2001) Evolution of the Aveiro Cretaceous aquifer (NW Portugal) during the Late Pleistocene and the present day: evidence from chemical and isotopic data. In: Edmunds, W.M. and Milne, C.J. (eds). Palaeowaters in coastal Europe: evolution of groundwater from the late Pleistocene. Geological Society of London Special Publications 189, 139–154.

Edmunds, WM (2001) Palaeowaters in European coastal aquifers – the goals and main conclusions of the PALAEAUX project. In: Edmunds and Milne (eds.) Palaeowaters of Coastal Europe; evolution of groundwater since the late Pleistocene. Geological Society, London, Special Publication 189, 1–16.

Edmunds, WM and Gaye, CB (1997) High nitrate baseline concentrations in groundwaters from the Sahel. Journal of Environmental Quality, 26, 1231–1239.

Edmunds, WM; Buckley, DK; Darling, WG; Milne, CJ; Smedley, PL and Williams, A (2001) Palaeowaters in the aquifers of the coastal regions of southern and eastern England. In: Edmunds and Milne (eds.) Palaeowaters of Coastal Europe; evolution of groundwater since the late Pleistocene. Geological Society, London, Special Publication 189, 71–92.

Edmunds, WM; Dodo, A; Djoret, D; Gasse, F; Gaye, CB; Goni, IB; Travi, Y; Zouari, K and Zuppi, GM (2004) Groundwater as an archive of climatic and environmental change – the PEP III traverse. (PEP III) Conference Aix-en Provence, August 2001. Kluwer

Gasse, F (2000) Hydrological changes in the African tropics since the Last Glacial Maximum. Quaternary Science Reviews 19: 189–211.

Hathaway, JC; Poag, CW and Valentine, PC (1979) US Geological core drilling on the Atlantic Shelf. Science, 206, 515–527.

Lehmann, BE; Love, A and Purtschert, R (2003) A comparison of groundwater dating with ^{81}Kr, ^{36}Cl and ^{4}He in four wells of the Great Artesian Basin, Australia.

Loosli, HH; Lehmann, B; Aeschbach-Hertig, Kipfer, R; Edmunds, WM; Eichinger, L; Rozanski, K; Stute, M and Vaikmae, R (1998) Tools used to study palaeoclimate help in water management. Eos, 79, 581–582.

Rozanski, K; Johnsen, SJ; Schotterer, U and Thompson, LG (1997) Reconstruction of past climates from stable isotope records of palaeo-precipitation preserved in continental archives. *Hydrological Science* **42**, 725–745.

Sturchio, NC; Du, X and Purtschert, R (2004) One million year old groundwater in the Sahara revealed by Krypton-81 and Chlorine-36. Geophysical Research Letters. 31, L05503, doi:10.1029/2003GL019234.

Vaikmäe, R; Edmunds, WM and Manzano, M (2001) Weichselian palaeoclimate and palaeoenvironment in Europe: background for palaeogroundwater formation. In: Edmunds, WM and Milne, CJ (eds.) Palaeowaters of Coastal Europe; evolution of groundwater since the late Pleistocene. Geological Society, London, Special Publication 189, 163–191.

KEYNOTE LECTURE

Groundwater flow system, subsidence and solute transport controls in the lacustrine aquitard of Mexico City

M.A. Ortega-Guerrero
Centro de Geociencias, Campus Juriquilla, Querétaro Universidad Nacional Autónoma de México

ABSTRACT: The Basin of Mexico has several flat plains formed by exceptionally porous (60–90%), clayey-size-rich lacustrine aquitard, overlying a highly productive regional aquifer that represents the main source of drinking water for about 20 million inhabitants in the Metropolitan Area of Mexico City. This paper reviews progress that has been made in field research pertaining to the role of the aquitard on the groundwater flow system, land subsidence, contaminant transport mechanisms and hydrochemical influence of leakage to the aquifer beneath, considering regional to local scales. Mexico City's aquitard plays an important role in the control of groundwater flow systems before pumping began in the lacustrine plains, and at present it has an important contribution with regional inflow through leakage and geochemical influence to underlying regional fresh water aquifer units, induced by pumping. Groundwater origin and flow, hydrochemical evolution and solute transport in this aquitard is also important for its evaluation for long-term hazardous containment and protective natural cover to underlying aquifer units. There are hydraulic and hydrochemical evidence of fracture flow in the aquitard; however, presence and maximum depth of active flow in deep, widely-spaced fractures on consolidation and solute transport is not yet well understood.

Keywords: Groundwater, aquitards, land subsidence, aquifer-aquitard interaction, Mexico City, groundwater flow system.

1 GENERAL SETTING

Mexico City is situated in the Basin of Mexico on a highly compressible, very porous (80–90%) lacustrine aquitard that overly highly productive aquifer units of both volcanic and sedimentary origin. Volcanic mountains close the basin (figure 1). This aquitard contributes with inflow through leakage to the underlying alluvial-pyroclastic-volcanic regional aquifer, from which about 50 m^3/s are obtained by pumping to supply water for the Metropolitan Area of Mexico City that concentrates some 20 million inhabitants and more than 30% of the nation's industry. This aquitard is important for the long-term management and protection of water resources in Mexico City and related urban areas. Figures 2a and 2b show the distribution of the hydrogeologic units and two cross sections that show the position of the lacustrine aquitard and the main aquifer units.

Severe land subsidence due to consolidation of the lacustrine aquitard is caused by groundwater extraction in the aquifer unit beneath, has resulted in restrictions on groundwater extraction in the core of Mexico City. This has led to large increases in extraction in the outlying lacustrine plains where satellite communities are rapidly expanding. One of these plains is the Chalco Plain where more of the studies presented in this paper have been carried out.

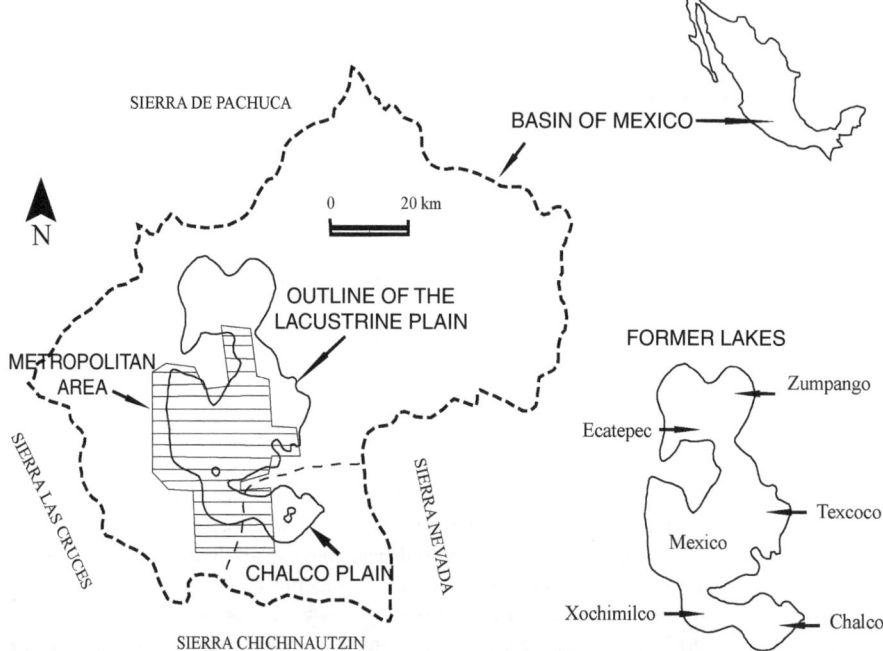

Figure 1. Location of the lacustrine areas within the Basin of Mexico and former lakes.

2 GROUNDWATER FLOW SYSTEM

2.1 *Natural manifestations of groundwater conditions*

Six ancient lakes existed in the region prior to 1789: Zumpango, Ecatepec and Texcoco in the lowest part of the basin contained brackish water (figure 1). The Aztecs built dikes to prevent the mixing of this water with fresh water of the lakes of Mexico, Xochimilco and Chalco during periods of high water levels (Durazo and Farvolden, 1989). At present, only part of Lake Texcoco artificially remains.

Groundwater discharge occurred at several places in the mountains at different elevations. From three centuries ago, important springs located in the margin of the lacustrine plane were used for the city municipal drinking water supply. Major springs also occur on the south side of sierras Chichinautzin and Las Cruces. These major springs, on opposite flanks of the mountain ranges, were assumed to be the intersection of the regional water-table with the ground surface, and the consequence of the interaction of all matrix factors such as hydraulic conductivity, distribution of hydrostratigraphic units and boundary conditions, and the main input function of groundwater recharge. Springs located in the foothills of Sierra Las Cruces caused flooding before the 1920's. Springs are not flowing at present near the lacustrine margin; they progressively disappear as a consequence of extensive groundwater extraction from the aquifer units beneath.

2.2 *Present manifestations of groundwater conditions*

Since the first deep boreholes were drilled in the core of Mexico City in the late 1880's, flowing artesian boreholes have been observed throughout the low topographic parts of the

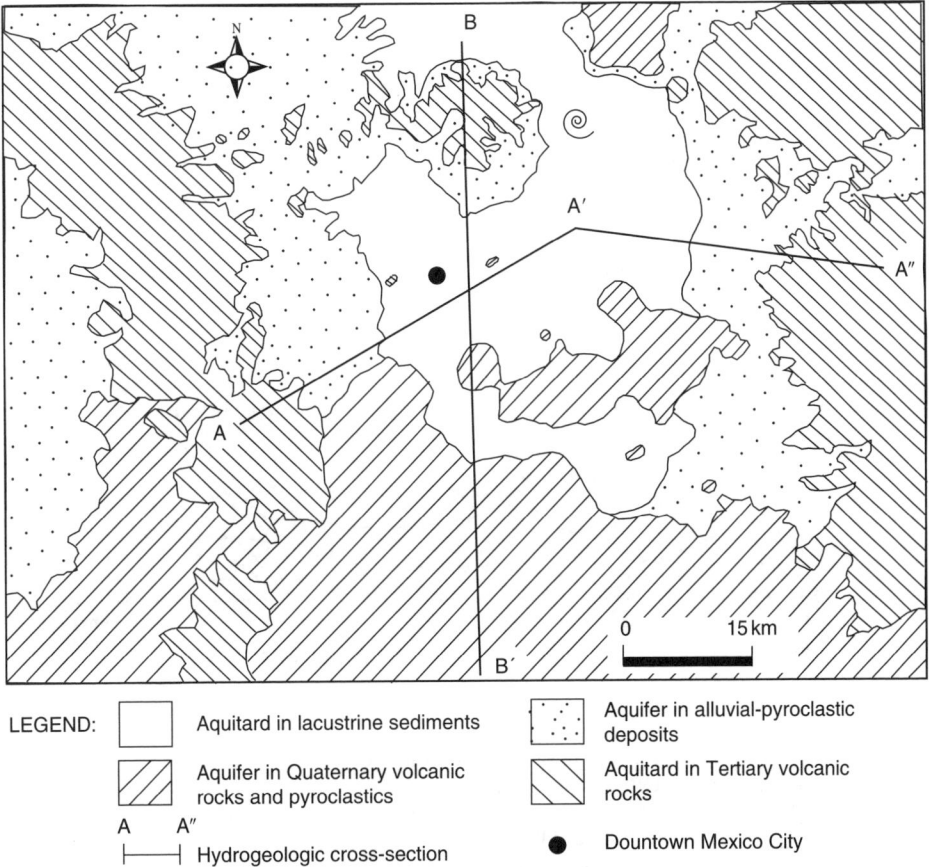

Figure 2. (a) Distribution of hydrogeologic units and (b) location of hydrogeologic cross-sections (After Ortega and Farvolden, 1989).

basin (Durazo and Farvolden, 1989). Multilevel piezometers installed to a maximum depth of 100 m by the Valley of Mexico Water Commission in the 1950's showed upward flow at nearly all locations. More than 10,000 boreholes exist now in the lacustrine area. This extensive groundwater extraction has caused the flow conditions to change; this component of the hydraulic gradient is now downward in most of the aquitard area.

Groundwater in the mountains and foothills in the basin of Mexico typically have low concentration of total dissolved solids and major ions, and no evidence of evaporation indicated by environmental isotopes $\delta^{18}O$ and δ^2H (Quijano, 1978). High altitude springs represent local discharge based also on environmental isotopes (Cortes and Farvolden, 1989).

In contrast, groundwater from piezometer nests in clayey deposits in the Chalco area shows a higher ionic concentration of chloride, bicarbonate and sodium in the upper 20 m than the water from the deeper nearby boreholes (200–400 m) (Ortega and Farvolden, 1989). Similar trends exist in the former Texcoco Lake where the shallow groundwater at 30 m is brine and shows the highest concentration of salts and the most extreme evidence of lacustrine evaporation effects in the Basin (Rudolph et al., 1991b; Ortega-Guerrero et al., 1997).

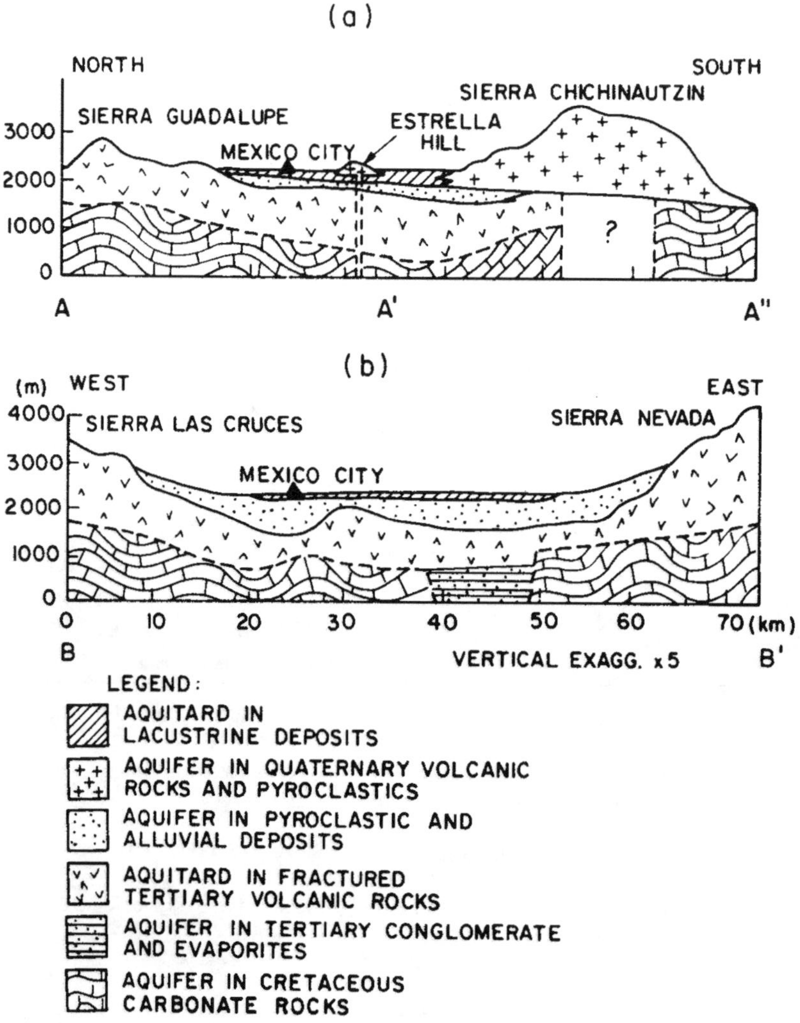

Figure 2. (b) Continued.

2.3 Groundwater flow cross-sectional modelling

Ortega and Farvolden (1989) presented a summary of the evidence of groundwater recharge and discharge in the Basin of Mexico, and used these evidences in combination with finite-element cross-sectional modelling to analyze the natural groundwater flow system and boundary conditions in the Basin of Mexico prior to extensive groundwater extraction. These authors found that the modelled flow patterns are consistent with the historical hydrologic records, piezometric characteristics and observed surface of groundwater features in the Basin of Mexico (figures 3a to 3d). Modelling results show groundwater recharge in the mountains to be 30–50% of the mean average precipitation. Higher and lower rates result in a flow regime that is not compatible with field observations. About 40–50% of the total discharge into the lacustrine plains was by upward flow through the lacustrine deposits. The position of the groundwater divides was also approximated. The flow system obtained

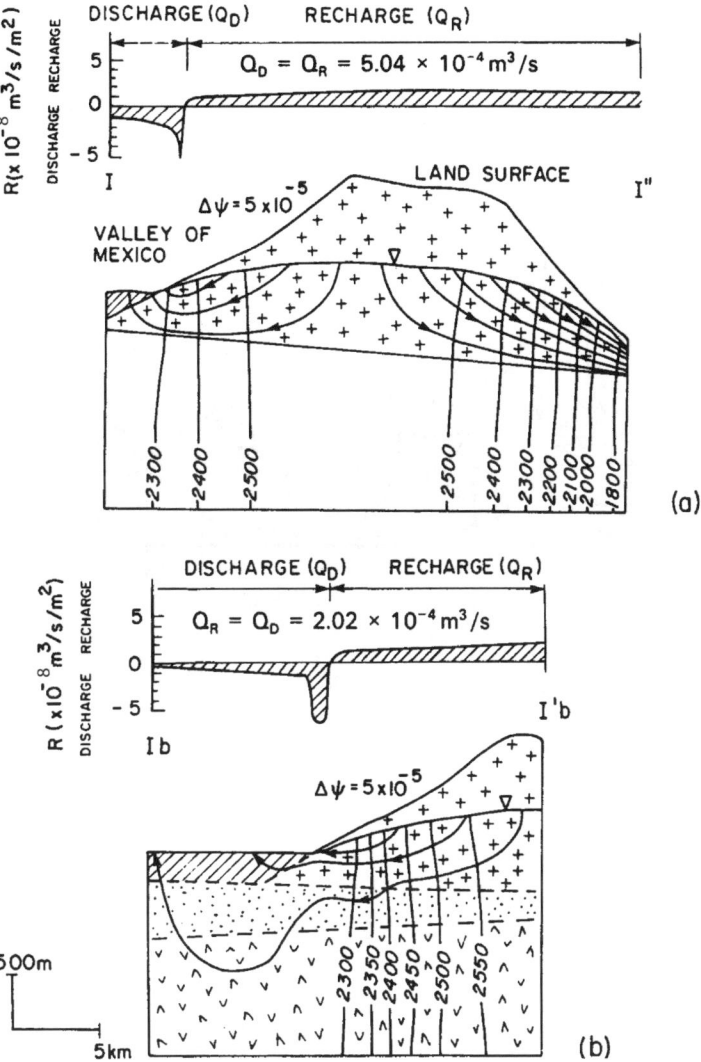

Figure 3. Groundwater flow system. Potential and streamfunction distribution: (a, b) Sierra Chichinautzin, (c) Sierra Las Cruces, and (d) Sierra Nevada. (After Ortega and Farvolden, 1989).

through numerical modelling is consistent with the classic gravitational groundwater flow as suggested by Tóth (1963, 1966) and by Freeze and Witherspoon (1967, 1968).

3 MECHANISMS CONTROLLING SUBSIDENCE

3.1 *Historical evolution of large scale consolidation in the Chalco Aquitard*

The extensive lacustrine Chalco Plain in the south-eastern part of Mexico City (figures 4a and 4b) is underlined by an aquitard up to 300 m thick composed of a layered sequence of very

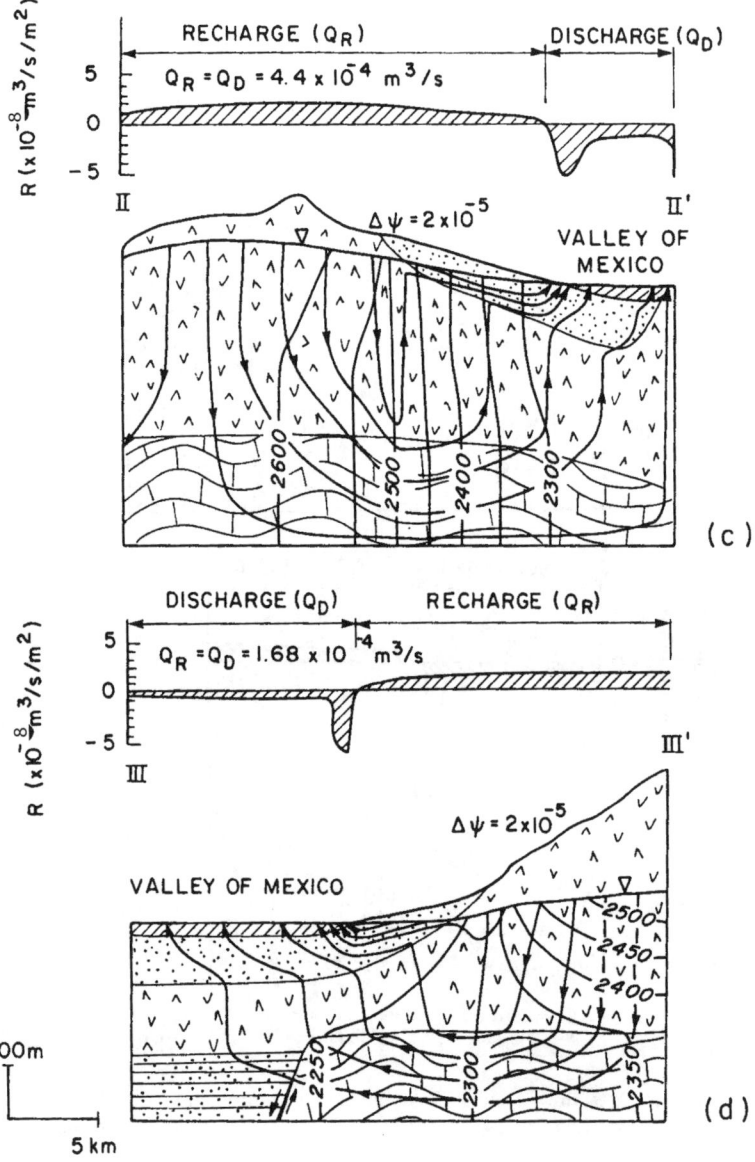

Figure 3 (continued).

porous fine-grained, organic-rich Quaternary deposits, with thin horizontal inter-stratified beds of pyroclastics ("*Capas Duras*"). The aquitard overlies a thick sequence of alluvial-pyroclastic material that forms a highly productive aquifer unit.

Groundwater flow conditions in the aquitard of the Chalco Plain show the area was a shallow lake until the 1940s when it was drained for agricultural use and human dwellings (Ortega-Guerrero *et al*., 1993). Historic information indicates that the Chalco Plain was under discharge conditions prior to the onset of heavy groundwater extraction from the aquifer unit beneath. This extraction reversed the hydraulic gradient throughout the full thickness of the aquitard in areas where the aquitard is thin (100 m), and recharging conditions now prevail.

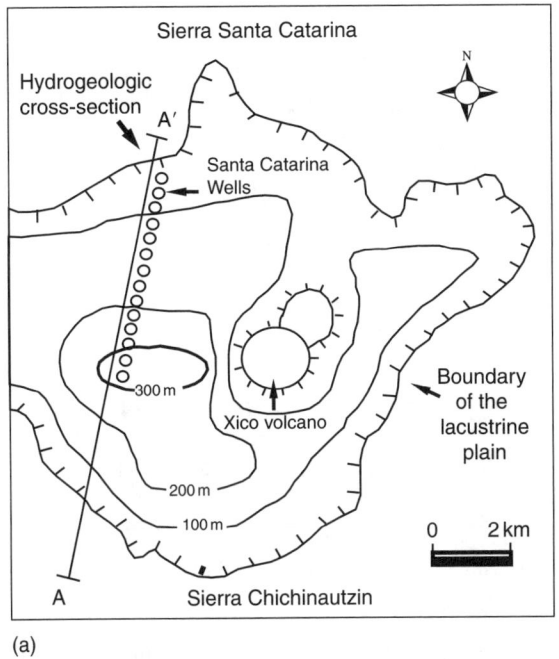

(a)

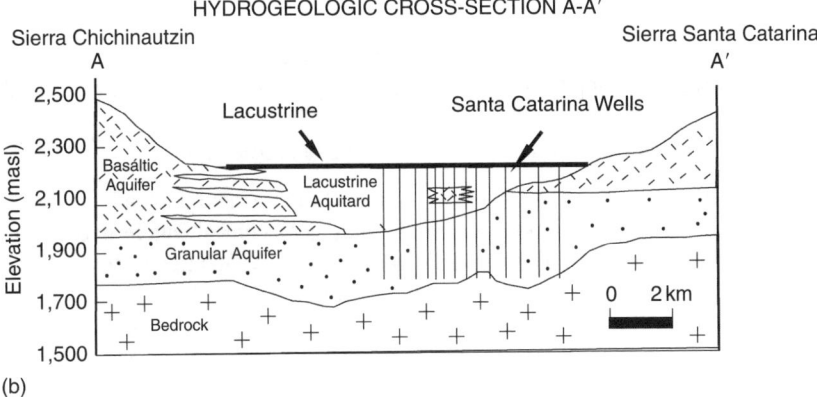

(b)

Figure 4. (a) Thickness of the lacustrine aquitard in the Chalco Plain. Location of the Santa Catarina Wells. (b) Hydrogeologic cross-section. (After Ortega-Guerrero et al., 1993).

Where the aquitard is thick, the hydraulic head data show a progressive decline with time even though the hydraulic gradient still indicates upward flow in at least the upper part of the lacustrine sequence.

In the early 1960s, when major groundwater extraction began beyond the periphery of the aquitard, and the onset in 1982 of heavy extraction from aquifer units beneath the aquitard, land surface subsided approximately 3 m. An additional subsidence of 2 m occurred between 1984 and 1989 (figure 5), causing a shallow lake to form and gradually expand. If the present rate of groundwater withdrawal from the Chalco Basin continues, total land subsidence in the middle of the plain will probably continue to a rate of about 0.4 m/year for many years, and could eventually reach a total subsidence of tens of meters in the thickest part of

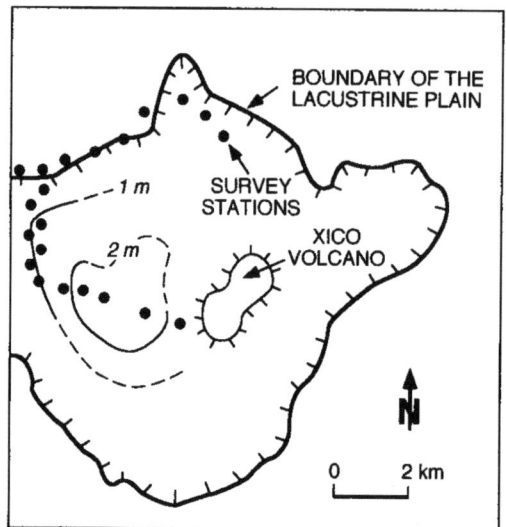

Figure 5. Cumulative subsidence over the Chalco Plain between 1984 and 1989 (After Ortega-Guerrero et al., 1993).

the Chalco Plain. Consequently, this area is susceptible to the highest potential land subsidence effects as a result of groundwater extraction anywhere in the basin. Land subsidence in the central part of the Chalco Basin has increased to 0.4 m/yr since 1984 and by 1991 total subsidence had reached 8 m. The rapid land subsidence in this area is causing the accumulation of meteoric waters during the rainy season resulting in extensive flooding of farmland.

3.2 Field instrumentation and numerical modelling of subsidence

The study by Ortega-Guerrero et al. (1999) demonstrated a methodology for combining hydraulic data from a network of monitoring boreholes, geotechnical data from core samples, and a compilation of historical information on land surface elevation to quantify groundwater flow and land subsidence phenomena within the rapidly subsiding Chalco Basin (figure 6). Then a One-dimensional mathematical model, which considers stress dependent parameters, was employed to develop predictions of future land subsidence under a range of extraction conditions. The model permits the hydraulic properties of the aquitard to vary as transient functions of hydraulic head and porosity. Simulations suggest that under current pumping rates, total land subsidence in the area of thickest lacustrine sediment will reach 15 m by the year 2010 (figures 7a and 7b). If pumping were reduced to the extent that further decline in the potentiometric surface is prevented, total maximum subsidence would be significantly less, about 10 m, and the rate would nearly cease by 2010 (figures 7a and 7b).

3.3 Scale dependence of the hydraulic parameters

Although the hydraulic diffusivity has much hydraulic and geotechnical importance, little is known about its magnitude and geologic controls at various spatial and time scales relevant to consolidation settlement. Different methods are used by Ortega-Guerrero (1996) to evaluate the hydraulic diffusivity (K/Ss) at four different scale volumes of sediment: a traditional odometer test, piezometer response tests, surface loading tests and modelling long term

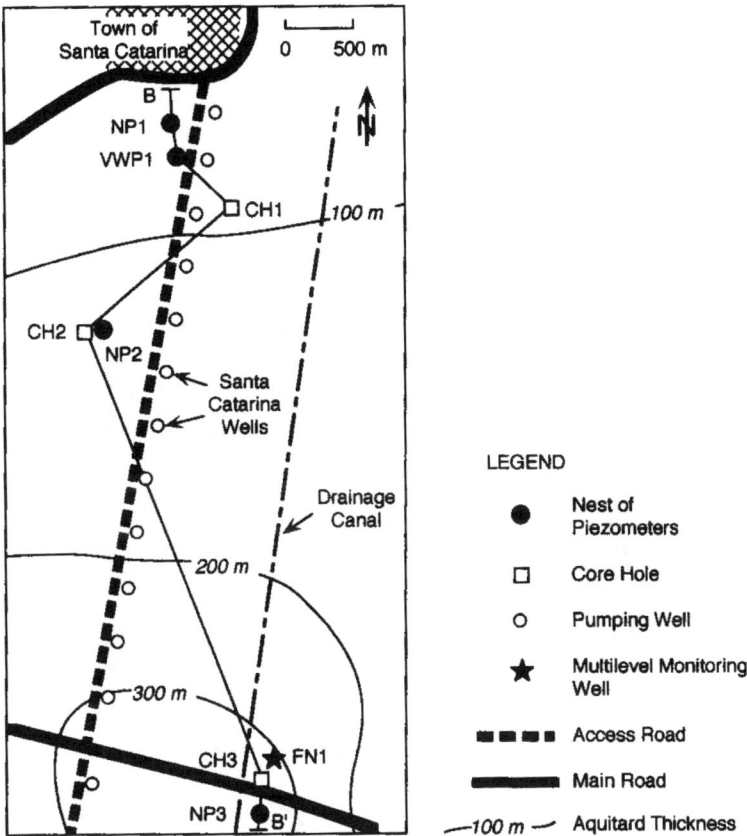

Figure 6. Groundwater monitoring sites and core holes used during the investigations relative to the Santa Catarina well field in the analysis of the evolution of land subsidence (After Ortega-Guerrero et al., 1999).

transient land subsidence due to groundwater extraction in an area where present subsidence rate is 0.40 m/year. The spatial scale of the measurements, range between 0.02 to 300 m, and the time scale between 24 hours to 30 years. Results show that the hydraulic diffusivity depends on the scale and time encompassed by each time of measurement. At the regional scale of the test, the hydraulic diffusivity increases, showing the increasing effect of discontinuities within the lacustrine sequence. When laboratory values are used in regional-scale subsidence models, results are unrealistic. The bulk hydraulic diffusivity is two orders of magnitude higher than the upper limit of the laboratory measurements (figure 8). Therefore, regional hydraulic diffusivity cannot be approximated from odometer tests and a range for this parameter has to be obtained based on the scale of application.

4 HYDRAULIC EVIDENCE OF FRACTURES

The aquitard's hydraulic conductivity (K′) is essential for evaluating the protection to the underlying aquifer unit and its hydraulic connection. The study by Vargas and

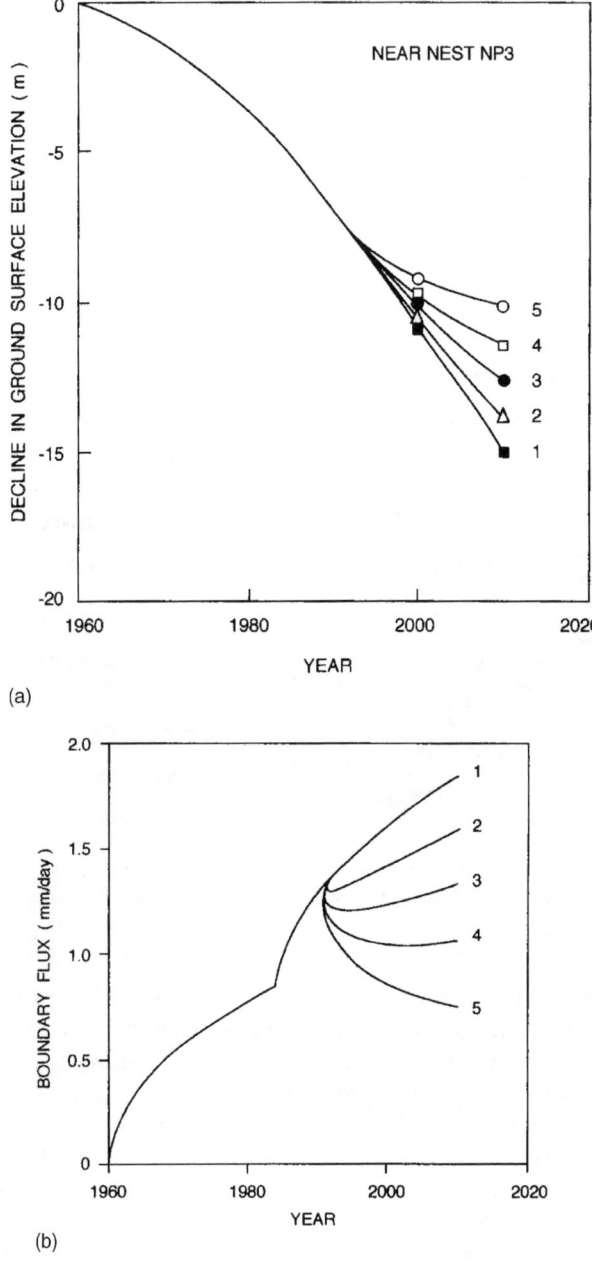

Figure 7. (a) Results of the numerical simulations of land subsidence. Predicted transient evolution of the total subsidence in the middle of the Chalco basin near monitoring location NP3 to the year 2010 if the rate of drawdown in the production aquifer is reduced by, 0% (case 1), 25% (case 2), 50% (case 3), 75% (case 4), and 100% (case 5). (After Ortega-Guerrero *et al.*, 1999). (b) Leakage flux entering the aquifer from numerical results. Predicted transient evolution of the total subsidence in the middle of the Chalco basin near monitoring location NP3 to the year 2010 if the rate of drawdown in the production aquifer is reduced by, 0% (case 1), 25% (case 2), 50% (case 3), 75% (case 4), and 100% (case 5). (After Ortega-Guerrero *et al.*, 1999).

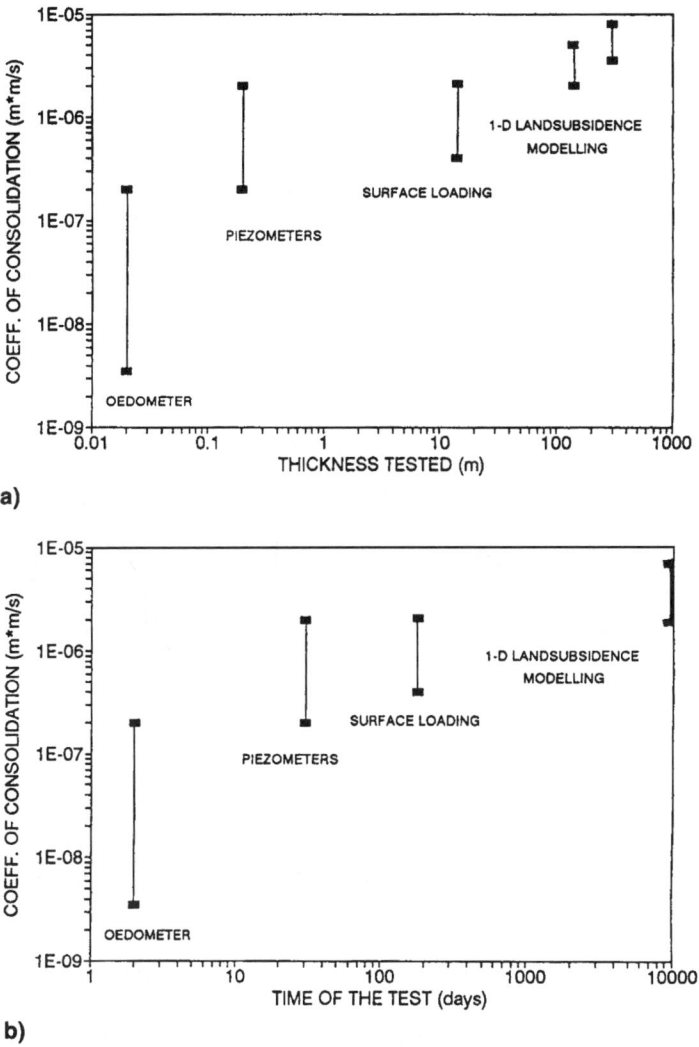

Figure 8. Hydraulic diffusivity (Coefficient of consolidation) as a function of: (a) the scale volume, and (b) time of test. (After Ortega, 1996).

Ortega-Guerrero (2004) analyzes the distribution and variation of K' in the lacustrine aquitard in the plains of Chalco, Texcoco and Mexico City (three of the six former lakes that existed in the Basin of Mexico), on the basis of 225 field-permeability tests, in nests of existing "drive point" – type piezometers located at depths of 2 to 85 m. Tests were interpreted by using the Hvorslev method and some by the Bower-Rice method. Results indicate that the hydraulic conductivity in the aquitards has a log-normal distribution and fit log-Gaussian regression models, with correlation coefficients above 97%. Dominant frequencies for K' in the Chalco and Texcoco plains range between 1E-09 and 1E-08 m/s, with population means of 1.19E-09 m/s and 1.7E-09 m/s respectively. Figure 9 shows the distribution frequencies for the hydraulic conductivity values obtained from the tests.

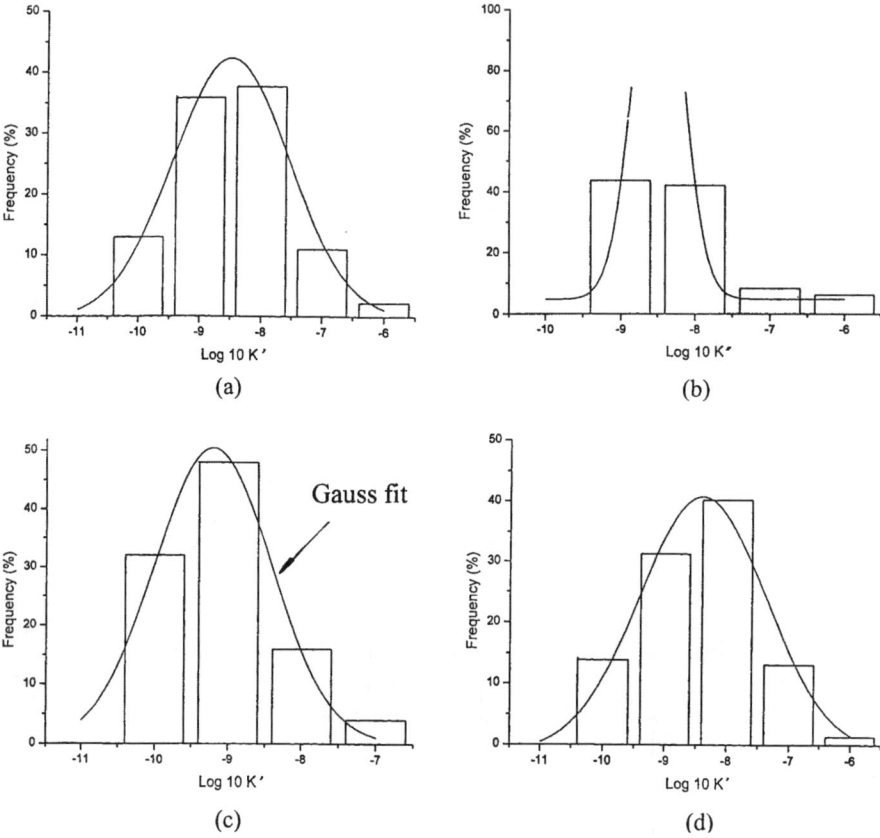

Figure 9. Distribution of frequencies for 225 hydraulic conductivity tests: (a) Chalco Plain, (b) Texcoco Plain, (c) Mexico Plain, and (d) the three zones. (After Vargas and Ortega, 2004).

In the Mexico City plain the population mean is near to one order of magnitude lower; $K' = 2.6\text{E-}10$ m/s. The increment of two orders of magnitude of the mean values with respect to the matrix conductivity is attributed to the presence of fractures in the upper part of the aquitard and perhaps in more deep levels, which suggests that the aquitard does not constitute a barrier to the migration of contaminants towards the aquifer unit beneath, and also that fracture flow might influence consolidation. This fracture hydraulic response is consistent with the findings of previous studies on solute migration in the aquitard (Rudolph et al., 1991; Ortega-Guerrero, 1993). Zawadsky (1996) also studied a regional scale fracture in the Chalco aquitard.

Results presented above show that widely spaced fractures might play an important role in the aquitard consolidation process. More recently, Aguilar-Perez et al. (2006) conducted an analysis of hydrodynamic fracturing near Mexico City through an integrated numerical analysis of the groundwater flow and geomechanical equations for land subsidence due to groundwater extraction of nine boreholes. Their results show that the critical pumping rate in the volcanic aquifer, between 420 l/s and 470 l/s, was exceeded since the beginning of the borehole field operation, which caused the mechanical failure of the overlying lacustrine materials and a series of fractures formed near the borehole field. The vertical ground deformation with time

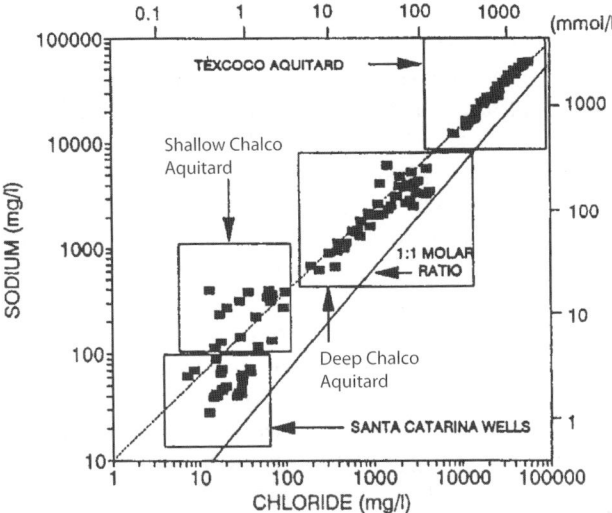

Figure 10. Relationship between log-chloride and log-sodium concentrations in groundwater samples from the Santa Catarina production boreholes and from piezometers in the aquitard (After Ortega-Guerrero et al., 1997). Data from the Texcoco aquitard is from Rudolph et al. (1991b).

cannot be reproduced in the numerical simulations with one set of parameters; two sets of parameters were needed to obtain a best fit, one for the 1960–1984 period, and another one for the 1985–1998. In September 1985 occurred one of the main earthquakes in Mexico City.

5 CHEMICAL INFLUENCE OF AQUITARD LEAKAGE AND CONTAMINANT TRANSPORT MECHANISMS

5.1 Origin of pore water and salinity in the aquitard

Pore water in much of the Chalco aquitard is saline; however, release of salts and other chemical constituents to the underlying aquifer unit has not yet significantly impaired the aquifer water quality (Ortega-Guerrero et al., 1993). Pore water in the extensive lacustrine aquitard of Mexico City ranges from brackish to highly saline and overlies a thick regional productive granular fresh water aquifer unit, which supplies water to the Metropolitan Area of Mexico City. Closed basin conditions developed since about 7,00,000 years ago and a series of six ancient connected lakes existed until 1789.

The study by Ortega-Guerrero et al. (1996) used stable isotopes $\delta^{18}O$ and δ^2H, major ion chemistry and geochemical modelling to investigate the origin and evolution of the highly saline-to-brine pore water aquitard. Groundwater samples were obtained from pumping boreholes in the aquifer and in the aquitard from nests of piezometers. Results show that the chemical patterns are consistent with the enrichment of $\delta^{18}O$ and δ^2H in the aquitard and follow the general evaporation trends developed for other closed or semi-closed basins in the world. These results demonstrated the role of evaporation as a dominant mechanism for the formation of high salinity pore water.

The concentrations of salts in the pore water aquitard are between three and four orders of magnitude higher than in the fresh water in the aquifer unit below. The chemical evolution of the aquitard pore water is associated with weathering reactions of sodium-plagioclase feldspars occurring in volcanic rocks in the recharge areas, where diluted water enters the regional groundwater flow system (Ortega, 1994). The lacustrine plains and periphery were extensive areas of groundwater discharge, which partially fed the former lakes in addition to runoff. Paleo-lake water underwent progressive evaporation that resulted in the present brackish sodium-bicarbonate aquitard pore water. Chemical patterns are consistent with the enrichment of $\delta^{18}O$ and $\delta^{2}H$ in the aquitard and follow general evaporation concepts developed for other closed or semi-closed basins in the world.

During evaporation anoxic conditions would prevail in the ancient lakes and cause strong sulphate and nitrate reduction in the presence of high organic matter content. The reduction of sulphate may also have contributed to the exceptionally high bicarbonate in the pore-water. Progressive evaporation of water in the aquifer unit beneath the aquitard was simulated with a geochemical model. After a concentration factor between 70 and 80, which represents evaporation of 98.6% to 98.75% of the original volume of water, the chemical trends for conservative ions observed in the aquitard pore water could be reproduced. These results clearly demonstrate the role of evaporation as the main mechanism responsible for the high salinity pore water in the Chalco aquitard (Ortega, 1994).

The persistence of paleo-lake water in the thick part of the aquitard indicates that upward advection was not sufficiently large throughout hundreds of thousands of years cause displacement of the original saline pore water. The deeper part of the delta of oxygen-18 and chloride profiles at one of the research sites where the aquitard id 140 m thick exhibit a gradual decline in concentration towards the bottom part, near the aquifer boundary. Numerical simulations of these profiles show that they are close to the steady–state position independently of of the time interval allowed fro downward diffusion from 1,00,000 to 10,00,000 years using moderate values of upward advection, controlled by the regional groundwater flow system, and downward diffusion coefficients (Ortega-Guerrero *et al.*, 1997).

Preliminary implications presented by Ortega (1994) show that reversal of hydraulic gradients in the aquitard due to heavy pumping for water supply is causing downward displacement of original pore water and mixing in the shallow part of the lacustrine sequence. In the deep part of the aquitard, increasing migration of paleo-lake saline water into the aquifer will occur and the input of adverse chemical constituents from the aquitard will increase in the future and eventually impair the quality of drinking water for Metropolitan Mexico City.

5.2 Contaminant transport mechanisms

There are different sources of solid and liquid waste from urban, industrial, agricultural and farm animal activities, which are disposed on the lacustrine aquitard in the Chalco Basin. This is critical in areas of the plain where the aquitard is thin. Ortiz (1996) and Leal (1997) studied the mechanisms of inorganic and organic transport underneath a canal site towards the underlying aquifer unit. Both authors concluded that micro-scale fractures (30–50 micrometers, spaced 1–2 m) in the shallow aquitard were controlling the migration of contaminants. Cervantes (1997) used enriched tritium profiles to study the effect of fracture flow near the canal site studied by Ortiz (1996) and Leal (1997), arrived to the same conclusions. Numerical modelling of natural solute transport in thick lacustrine zones in Texcoco and Chalco also show micro-scale fracture controls on flow (Rudolph *et al.*, 1991b; Ortega, 1994).

6 CONCLUSIONS

Mexico City's aquitard has played an import role in different hydrogeologic aspects within the central and southern part of the Basin of Mexico. The distribution and hydraulic conductivity of the aquitard controlled the groundwater flow system in the basin, the amount of discharge at the edges of the plain, and the distribution of evaporated saline pore water towards the aquitard. As a consequence of extensive groundwater extraction, hydraulic gradients have reversed and they are now downwards where the aquitard is less than 100 m.

Severe land subsidence, due to consolidation of the lacustrine aquitard caused by groundwater extraction has reached near 10 m in the middle of the Chalco Plain causing the development of regional fractures in areas where high heterogeneity exist.

There are field evidences of micro-scale fracture controls on groundwater flow, hydrochemistry and contaminant transport in the aquitard. Results show that the aquitard does not represent a barrier for contaminants disposed on ground surface in areas where it is less than 20 m. The widely spaced fractures might play an important role in the aquitard consolidation process and solute transport; however, there are instrument limitations at present to evaluate these processes.

ACKNOWLEDGEMENTS

Most of the projects referred in this paper were funded by NSERC operational projects to John A. Cherry from the University of Waterloo and operational grants by CONACYT and the National University of Mexico (UNAM) to Adrian Ortega. The author wish to thank J. A. Cherry, R.N. Farvolden (+), D. Rudolph and R. Aravena from the University of Waterloo for the advice provided during many critical stages of the aquitard research, and to R. Ingleton and Martin Vallejo for field assistance and instrumentation. Thanks to R. Ortega for his continuous support.

REFERENCES

Aguilar-Pérez, L.A., Ortega-Guerrero, M.A., Hubp, J.L., and Ortiz, Z.D.C (2006) Análisis numérico acoplado de los desplazamientos verticales y generación de fracturas por extracción de agua subterránea en las proximidades de la Ciudad de México. [Coupled numerical analysis of vertical displacements and fracture generation induced by aquifer pumping near Mexico City]. Revista Mexicana de Ciencias Geológicas. 23(3), pp. 247–261.

Bribiesca-Castrejón, J.L (1960) Hidrología Histórica del Valle de México [Historic hydrology of the Valley of Mexico]. Ingeniería Hidráulica, v xiv, n. 1, México, pp. 107–125.

Carrillo, N (1947) Influence of Artesian Wells in the Sinking of Mexico City. Comisión Impulsora y Coordinadora de la Investigación Científica, Anuario 47. In: Volumen Nabor Carrillo, pp. 7–14, Secretaría de Hacienda y Crédito Público, México, 1969.

Cervantes, M.A (1996) Tritio como indicador de los mecanismos de transporte al Noreste del acuitardo lacustre de Chalco. [Tritium as indicator of groundwater age in the Chalco aquitard] MSc Thesis, Posgrado de Ingeniería Ambiental, Facultad de Ingeniería, UNAM.

Cortes, A. and Farvolden, R.N (1989) Isotope studies of precipitation and groundwater in the Sierra Las Cruces Mexico. Journal of Hydrology, 107:147–153.

Durazo J. and Farvolden R.N (1989) The groundwater regime of the Valley of Mexico from historic evidence and field observations. Journal of Hydrology 112:171–190.

Freeze, A. and Cherry, J. A (1979) Groundwater. Prentice Hall, USA, 604 p.

Freeze, A., and Witherspoon, P (1967) Theoretical analysis of regional groundwater flow: 2. Effect of water table configuration and subsurface permeability variation. Water Resources Res., 3, pp. 623–634.

Freeze, A. and Witherspoon, P (1968) Theoretical analysis of regional groundwater flow: 3. Quantitative interpretations. Water Resources Res., 4, pp. 581–590.

Hvorslev M.J (1951) Time lag and soil permeability in groundwater observations. Waterways Experiment Station Corps. Of Engineers, U. S. Army, Vicksburg, Missisipi, Bol. 36, 50 p.

INEGI e INE (2000) Indicadores de Desarrollo Sustentable en México [Indicators for sustainable development in Mexico] Instituto Nacional de Estadística, Geografía e Informática y el Instituto de Ecología, 5 de Junio de 2000.

Leal, B. R.M (1997) Migración de compuestos orgánicos en un sitio de canal de agues residuales hacia el acuífero subyacente, Chalco, México. [Migration of organic compounds at a canal site to the underlying aquifer, Chalco, Mexico]. MSc Thesis, Posgrado de Ingeniería Ambiental, Facultad de Ingeniería, UNAM.

Marsal R.J (1969) Development of a lake by pumping-induced consolidation of soft clays. In Volumen Nabor Carrillo, Secretaría de Hacienda y Crédito Público, México, 11:229–266.

Marsal R.J. and Mazari M (1990) Desarrollo de la mecánica de suelos en la Ciudad de México. [Evolution of soil mechanics in Mexico City]. In: El Subsuelo de la Cuenca del Valle de México y su Relación con la Ingeniería de Cimentaciones a Cinco Años del Sismo, Sociedad Mexicana de Mecánica de Suelos, México 1:3–24.

Ortega, G.M.A (1994) Origin and migration of pore water and salinity in the consolidating Chalco aquitard, near Mexico City, PhD dissertation, University of Waterloo, Ontario, Canada.

Ortega, G.M.A. and Farvolden, R.N (1989) Computer analysis of regional groundwater flow and boundary conditions in the Basin of Mexico. Journal of Hydrology 110:271–294.

Ortega-Guerrero, M.A (1996) Variability of the coefficient of consolidation of the Mexico City clayey sediments on spatial and time scales. Bulletin of the International Association of Engineering Geology, Paris 54:125–135.

Ortega-Guerrero, M.A., Cherry, J.A., and Rudolph, D.L (1993) Large-scale aquitard consolidation near Mexico City. Ground Water 31(5):707–718.

Ortega-Guerrero, M.A., Cherry, J.A. and Aravena, R (1996) Origin of pore water and salinity in the México City aquitard. Journal of Hydrology.

Ortega-Guerrero, M.A., Rudolph, D.L., and Cherry, J.A (1999) Analysis of long-term land subsidence near Mexico City: Field investigations and predictive modeling. Water Resources Research 25(11):3327–3341.

Ortiz, Z. D.C. Migración de contaminantes inorgánicos derivados de un canal de desechos urbanos e industrials hacia el acuífero subyacente, Chalco, México. [Migration of inorganic contaminants from a urban sewage canal to the underlying aquifer] MSc Thesis, Posgrado de Ingeniería Ambiental, Facultad de Ingeniería, UNAM.

Rudolph, D.L. and Frind, E.O (1991a) Hydraulic response of highly compressible aquitards during consolidation. Water Resources Research 27(1):17–30.

Rudolph, D.L., Cherry, J.A., and Farvolden, R.N (1991b) Groundwater flow and solute transport in fracture lacustrine clay near Mexico City. Water Resources Research 27(9):2187–2201.

Tóth, J (1963) A theoretical analysis of groundwater flow in small basins. J. Geophys. Res., 68(16): 4795–4812.

Tóth, J (1966) Mapping and interpretation of field phenomena for groundwater reconnaissance in a praire environment, Alberta, Canada. Bull. Int. Assoc. Sci. Hydrol., 11(2):1–49.

Quijano, L (1978) Comentario sobre el muestreo de las aguas subterráneas de la zona de Texcoco. Int. At. Energy Agency (IAEA) Sec. Agric. Rec. Hid.

Zawadsky, A (1996) Investigations regarding the origin and hydrologic activity of a large-sediment-filled fissure in fractured lacustrine clay near Mexico City. MSc. dissertation, University of Waterloo, Ontario, Canada.

KEYNOTE LECTURE

Groundwater flow system response in thick aquifer units: theory and practice in Mexico

J.J. Carrillo-Rivera[1] and A. Cardona[2]
[1]*Instituto de Geografía, UNAM, CU, Coyoacán, México, DF*
[2]*Facultad de Ingeniería, UASLP, Zona Universitaria, San Luis Potosí, México*

ABSTRACT: Understanding before analysis, is a motto that requires further attention in groundwater studies as a general practice in Mexico and perhaps in other countries in the world. The perception achieved of groundwater functioning in shallow thin aquifers has provided a good base to define a problem and to propose agreeable solutions where a groundwater response is evaluated. Standard application of available evaluation tools that were usually devised for isothermal systems requires specific considerations to be practiced in systems where groundwater of different chemical composition and temperature flows in a stratified fashion and points of observation have commonly been disturbed by withdrawal resulting in observed water levels that fail to represent actual hydraulic potentials. Changes in density influenced by chemistry and temperature of outflowing water at a borehole, might mislead the presence of semi-confining conditions through the variation of drawdown with time (s-t) in pumping-test data interpretation. Information related to the geological framework becomes paramount in regions where thick aquifer units (*ie*, $\geqslant 1,500$ m) are the main control of groundwater flowing in local intermediate and regional systems. Two main issues araise in the analysis of three thick regional aquifers in central México: (1) The position of the basement rock and (2) groundwater chemistry changes and density effects, to define and develop water sources based on an adequate analysis of pumping-test results interpretation that uses correct hydraulic potential computations and that are adequately included in the the modelling of flow and of course in the expected water quality in withdrawal boreholes, largely in the expected transient groundwater flow systems that may develop, all of which become more important in thick aquifer units.

Keywords: regional flow, groundwater quality change, basement rock, thick aquifer unit, Mexico

1 INTRODUCTION

Groundwater evaluation studies involve several well known procedures in hydrogeological sciences, such as the application of the continuity equation under specific boundary conditions according to the characteristics of the flow domain. The continuity equation is applied on a flow domain where groundwater flow is assumed to take place in a specific time lapse and in a three dimension field. The chemical and physical characteristics of the groundwater flow provide information on the interaction between groundwater and the geological material that is occupied by an active flow system. In a strict sense, the vertical limits that bound and control saturated groundwater flow are referred to those placed by the geological framework where the bottom part is the basement rock and the top part of the

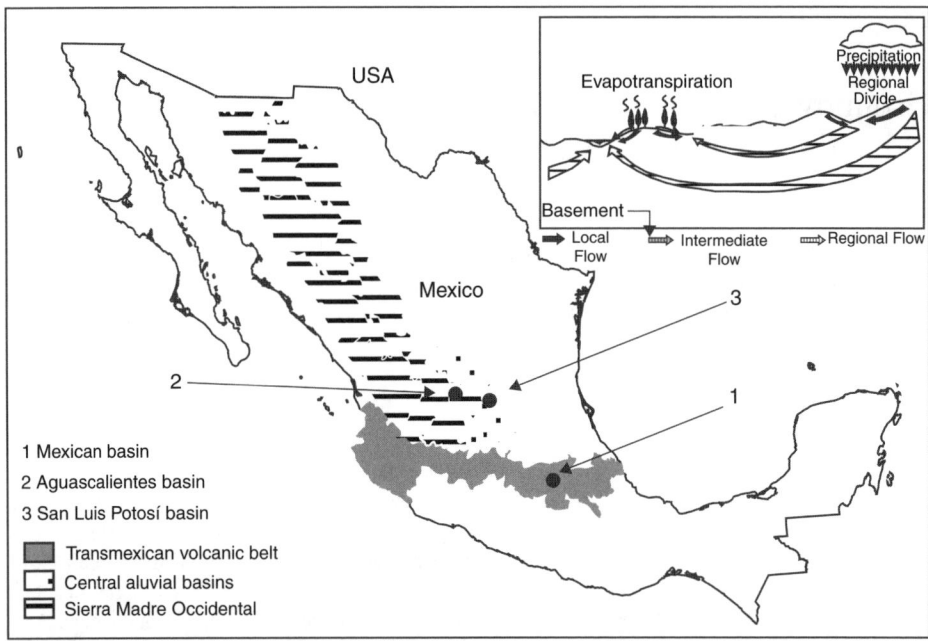

Figure 1. Location of study areas and their physiographic provinces, inset a sketch of the groundwater flow system adapted and simplified from Tóth (1999).

groundwater flow domain is represented by the potentiometric surface; lateral boundaries are those of the regional flow path travelling from the top of the foremost watershed (see sketch inset in figure 1) to its discharge area.

The physical and chemical characteristics that the flow scale (local o regional) include on groundwater become important evidence in groundwater studies, they suggest the apparent dimensions of the basin where groundwater flow is taking place. The practical aspects on how elements of the flow are defined through field measurements are a special topic, where the geological framework becomes an important issue. A shallow (few tens of meters) basement rock could imply that a borehole may fully penetrate the aquifer material; the path length and the temperature difference in the vertical is not so contrasting as that imposed by a lengthier path and a geothermal gradient resulting from a thick aquifer unit. Groundwater movement to a borehole in a shallow aquifer unit away from the recharge and discharge zones is, in a practical sense, controlled mainly by horizontal flow. The nature of flow through pores or fractures and the representative elementary volume provide an important reference on the local scale to understand groundwater flow at regional scale. In thick aquifer units boreholes are often partially penetrating, often they bottom above intermediate and regional flows that are able to travel several hundred of meters, where withdrawal may induce deep flow producing higher temperature water than that of local systems. Mexico, as well as many other territories in the world, has fractured aquifer units with more that 1,500 m of thickness; some examples have been identified in the physiographic provinces of the Central Alluvial Basins (Carrillo-Rivera et al., 1996), the Mexican Transvolcanic Belt (Ortega and Farvolden, 1989; Edmunds et al., 2002), the Sierra Madre Occidental (Carrillo-Rivera et al., 2001) and territories in The Baja California peninsula (Carrillo-Rivera, 2000), all in which withdrawal boreholes disturb the field of flow by obtaining groundwater at an average depth of about 300 m. These partial penetration conditions imply that production boreholes usually produce a three dimensional (3D) radial flow to their screened section inducing lateral water flow as well as flow from beneath in different proportions depending mainly on: the hydraulic conductivity (K) field distribution and the hydraulic potential. The latter is influenced by an interaction among drawdown, thickness of the shallow local or intermediate flows (cold) groundwater, and the density variation contrast between the cold and the deep regional groundwater flow (usually thermal). The objective of this review is to draw attention on the significance of groundwater extraction in these particular hydrogeological conditions, discussing the specific response in both water quality and hydraulic potential; information that could be used to highlight the presence and importance of regional groundwater flow systems and help to draw attention to practical issues such as pumping-test interpretation and chemical evolution in time and space to reach an understanding of the existing groundwater flow systems related to thick aquifer units. It is expected that the practical applications of these results would improve the evaluation, management and conservation policies of groundwater resources in central México.

2 THEORETICAL BASIS AND CONSIDERATIONS

Four hydrogeologic methods, which are commonly applied to the evaluation of groundwater resources in Mexico, are addressed to discuss their potential and limitations when they are apply to thick aquifer units:

2.1 Pumping tests

Standard hydrogeological procedures in Mexico involved in the use of standard pumping-test (constant yield, s-t variation, borehole construction design and encountered lithology) are commonly interpreted by analytical methods according to the rate of s-t response. Results are commonly analysed to obtain transmissivity (T) and storage coefficient (S) of aquifer material according to theoretical conditions defined in the literature in agreement with a defined hydraulic response: confined (Theis, 1935), semi-confined (Hantush, 1952), water-table (Boulton, 1954) or fractured media (Boulton and Streltslova, 1978). Rathod and Rushton (1991) presented an alternative pumping-test analysis method of interpretation that incorporates all of the restrictions (negligible borehole radius, infinite aquifer extension, constant withdrawal yield, fully penetration, isotropy, delay gravity drainage, negligible well-losses, radial horizontal flow) as established in the above analytical solutions. Although the first four restrictions may be clearly defined based on direct field evidence, the application of analytical methods expects that the prevailing flow incorporates restrictions such as horizontal flow with a resulting drawdown that corresponds to a change in storage.

The aquifer thickness crossed by a borehole becomes paramount; should the borehole partially penetrates an aquifer, the larger the withdrawal rate and the larger of vertical hydraulic conductivity (K_v) as related to the horizontal hydraulic conductivity (K_h) the more important the vertical components of flow become. The fractured nature of the lithology crossed by a borehole provides an insight to the presence of flow through preferential paths such as along fault planes and related fractured features. In contrast, an aquifer unit with granular porosity that is not affected by any fracture structure could minimize the vertical components of flow by partial penetration effect (mainly when the value of K_h is larger than K_v). The presence of an aquifer unit overlied by a confining aquitard unit with low vertical hydraulic conductivity (K'_v) provides a leaky response to a withdrawal borehole when both units are hydraulically connected (*ie*, the potentiometric surface is above the base of the aquitard). This mechanism of vertical flow has been defined in the literature (Hantush, 1956) and could be confirmed when such leaky inflow is identified by its chemical signature.

When horizontal flow towards a withdrawal borehole prevails during a pumping-test, the upper limit of the tested volume of an aquifer unit is represented by the cone of depression in the potentiometric surface; the lower limit is the basement rock unit that is often represented by a plane bounded by the depth of the borehole. When the influence from neighbouring production boreholes as well as streams, canals and water bodies in the zone of the cone of depression are absent, observed *s-t* results are considered to be influenced only by the withdrawal of the borehole.

In general, a reduction in the rate of drawdown with withdrawal time (*s-t*) could result from four basic scenarios, all of which depict contrasting groundwater flow conditions: (*i*) inflow from a clayey strata in the tested site that through a leaky effect functions as a semi-confining layer (Hantush, 1956); (*ii*) withdrawal in a borehole located within the influence of the tested borehole is halted; (*iii*) inflow from a water body, or stream, hydraulically connected to the tested site; and (*iv*) the borehole penetrates the upper part of a thick aquifer unit, at least two groundwater flow systems are found and thermal water is induced from depth (Carrillo-Rivera *et al.*, 1996).

When any of such scenarios is met in a test, the observed evolution rate of the dynamic water level is reduced, producing often difficulties in the appropriate interpretation of the particular flow conditions. Field *s-t* measurement during a pumping-test could be more adequately interpreted should field physicochemical parameters are collected contemporaneously during the test, and assisted by particular water quality analyses (*ie*, trace elements, isotopes found in the geological media of interest).

An important feature of pumping-test procedure is to obtain the drawdown variation with withdrawal (or recovery) time, which is usually achieved by comparing the static water level with the dynamic water levels observed during the test in terms of "*depth to the water level*"; it is usually considered that the difference in hydraulic potential (drawdown) is equivalent to the difference in the measured water levels. This general operation is applicable in thin and shallow aquifer units where obtained groundwater has constant temperature and salinity content, as not to produce density difference effects of more than 5–10%. This means that the flow of groundwater to the withdrawal zone of the borehole is produced solely by the hydraulic gradient resulted from the lowering of the water level by the pumping yield in the tested borehole.

2.2 Hydraulic Potential

When groundwater is obtained in a borehole that partially penetrates an aquifer unit that has different flow systems in which water temperature increases with depth, this induces thermal water to upper aquifer levels; here, the determination of the hydraulic potential difference (drawdown) requires adjustment other than computing water level differences with the evolution of pumping time. If natural groundwater velocity is considered to be negligible, the hydraulic potential depends on the elevation of the observation point above a reference level and the weight of the water column as defined by Hubbert (1940) for fluids with a variable density,

$$\Phi = zg + \int_{p_o}^{p} \frac{dp}{\rho};$$ (Equation 1)

Where,

Φ hydraulic potential
z elevation of observation point, in relation to reference level
g acceleration of gravity
p pressure of water
p_o atmospheric pressure
ρ fluid density.

Equation 1 could be interpreted that the hydraulic potential (and drawdown) may be obtained if a pressure device is placed at the base of the aquifer (just on the surface of the basement rock unit below the tested borehole) to measure the pressure of the water column. Any change in the hydraulic potential (drawdown) will be reflected as a measure of the difference in pressure of the water column (instead of the difference in water level elevation), which will be given by the change in water density resulting from the continuous inflow of water with contrasting amount of dissolved salts and (usualy) higher water temperature than that initially obtained, generating an up-conning of groundwater flow with time.

The up-conning of groundwater flow is a phenomena of similar nature to that observed in the classical case of seawater intrusion; the analogous density difference (about 0.02 gr/cm^3) caused by salinity contrast between sea water and fresh water is the major gradient that influence seawater to move inland into the fresh water flow located above. In the case of cold and thermal water the density difference is quite comparable; for instance shallow water with 25°C and thermal water with 75°C have a density value of 0.9970 gr/cm^3 and 0.9765 gr/cm^3, respectively. Analogously, when the reduction in thickness of the cold-water body is reduced by withdrawal, the steady state buoyancy condition of the stratified cold-water flow is influenced by a resulting displacement of an ascending thermal-flow from beneath by the reduction in head-potential in the shallow cold-flow. The rate of ascending thermal water flow will produce a continuous change in the field of flow and the physical and chemical quality of the obtained flow. In an aquifer column beneath the withdrawal borehole the replacement of cold water by thermal water (figure 2) implies that the pressure as measured at the bottom of the aquifer (top of basement rock) will require a higher thermal water column than when cold water was present; this response in a thick (>1,500 m) aquifer media may be manifested as a rise in the water level. In this regard a correction for the hydraulic potential could need a subtraction of up to several meters out of the measured depth to the water level (Kaweki, 1995; Hergt and Carrillo-Rivera, 2004). This implies that the movement of groundwater into a pumping borehole has important vertical components of flow. Borehole 1 in figure 2 might indicate a case where beneath there are two groundwater flows, one cold overlaying a second one themal in nature; its hydraulic head as measured at sea level is of 1,973 m; in nearby borehole 2 the recorded level several hours after pumping stopped indicated a water-table elevation some 3.0 m above the level of borehole 1, suggesting growndwater moves from

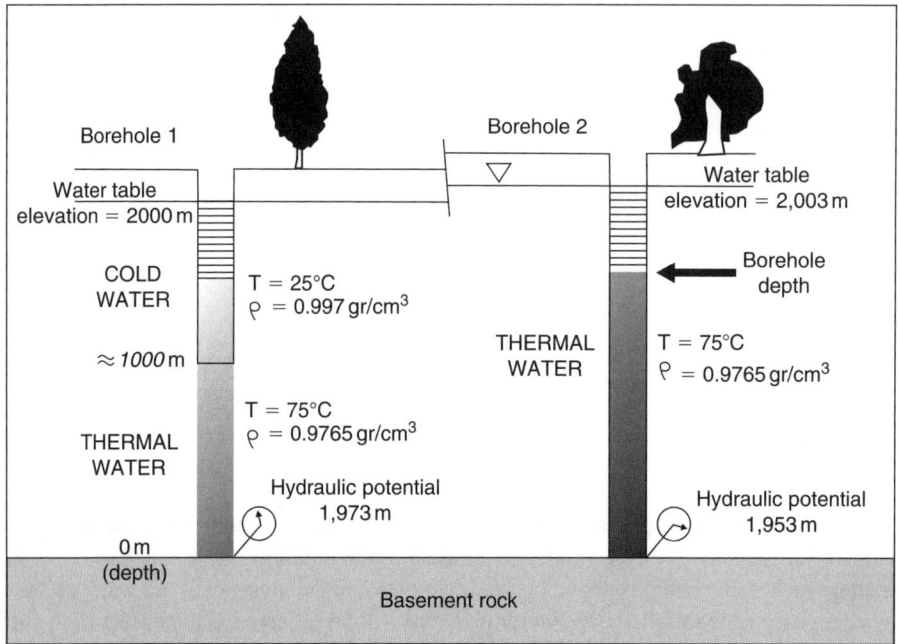

Figure 2. Difference in hydraulic potential due to the inducement of groundwater with contrasting density.

borehole 2, to borehole 1. However, the hydraulic head below borehole 2 is of 1,953 m indicating that the direction of groundwater flow is from borehole 1 to borehole 2.

Direct measurement of temperature (or salinity) with depth by means of logging could prove useful; however, such procedure is often difficult to carry out to the total depth of the aquifer unit. Here, the application of geothermometers might prove as a valuable tool to estimate the minimum equilibrium groundwater temperature at depth of the thermal flow. The use of such value with the geothermal gradient provides an adequate depth estimate of groundwater flow which may be cross referenced with geological records and geophysical data in regard to aquifer thickness.

2.3 Modelling groundwater

The three-dimensional (*3D*) movement of groundwater as incorporated in the partial differential equation (equation 2) is solved through computer packages as Modflow (McDonald and Harbough, 1988), wich is the official model in Canada and the USA and the more popular comercial model used in Mexico and perhaps in other countries.

$$\frac{\partial}{\partial x}\left(K_{xx}\frac{\partial h}{\partial x}\right)+\frac{\partial}{\partial y}\left(K_{yy}\frac{\partial h}{\partial y}\right)+\frac{\partial}{\partial z}\left(K_{zz}\frac{\partial h}{\partial z}\right)-W=S_s\frac{\partial h}{\partial t} \qquad \text{(Equation 2)}$$

This model incorporates constant density as a restriction (*ie*, isothermal and constant salinity in all direction and time). In general, the specific storage coefficient (S_s) and the hydraulic conductivity tensor (K_{xx}, K_{yy}, K_{zz}) may be a function of space; the volumetric flux (W) may be a function of space and time. The flow thus developed is under non-equilibrium conditions in a heterogeneous and anisotropic medium, where the principal axes of the hydraulic conductivity are meant to be aligned with coordinate directions. A solution to equation 2 incorporates values of the hydraulic potential in time and space based on the initial hydraulic potential conditions. Consequently, in practical applications where regional-thermal groundwater flow systems exist, if only partial penetrating active boreholes are used to obtain hydraulic potential (Φ_{field}) data, the larger the thickness of the aquifer unit under study, the more the values of hydraulic potential (Φ_{model}) resulting from the application of a model may fail to represent those observed in field conditions, as they are affected by the inflow of thermal water that may change the expected theoretical value of the hydraulic potential.

The three-dimensional movement of groundwater has been considered in most standard modelling procedures as Modflow (McDonald and Harbough, 1988). This is a widely used method that incorporates the variation in time of the hydraulic potential resulting from a volumetric flux per unit volume (represented by a source or sink of water) that is in agreement with the variation in the corresponding hydraulic potential developed in the x,y,z coordinates, which in turn is influenced by the hydraulic conductivity in each of those directions according to the specific storage coefficient of the aquifer unit that is expected to be a porous media. Although the aquifer material in a study area could be fractured, in many cases, the elementary representative volume maybe invoked to use this media as equivalent to a porous media due to a scale effect when modelling is applied in small scale (regional) problems (Bear, 1972).

2.4 Groundwater flow systems considerations

An understanding of groundwater functioning could be achieved estimating flow paths in both horizontal and vertical planes, with recharge and discharge areas resulting. In general three main groundwater flow systems are to be defined within the topography and geological framework: local, intermediate and regional (Tóth, 1999). A hilly topography, representative of central Mexico (inset in figure 1), may produce various local systems; where, part of the infiltrated water that enters, also leaves the same valley. In some cases, part of the recharged water may discharge in another river channel located at a lower topographic level, implying an intermediate system. A regional system travels to the deepest part of the basin, and it developes from the highest groundwater-divide to the lowest discharge area. Water belonging to a regional flow may have higher temperature than a local (or intermediate) flow which has travelled along a shallow depth. These steady-state flows in their natural geological media keep their paths separated. The natural hydrological conditions and the chemical, physical and biological aspects in a recharge area are reflected in a particular soil and vegetation cover; and they contrast with those in a discharge area. Recharge and discharge areas are strictly controlled by vertical flow with a downward and upward groundwater movement, respectively.

The *flow system theory* as develop by Tóth (1962, 1963, 1995, 1999) incorporates the major natural processes linked to groundwater flow that are consistent among themselves and that may be clearly established, and confirmed, by their co-dependence with the interpretation with tools belonging to other disciplines, such as geological framework, topographic elevation, basement rock position, water chemistry evolution, heat flow transfer, soil characteristics, vegetation cover, and presence or absence of surface water features. Due to the review nature of this paper, only basics of groundwater flow systems characterization were used, such as temperature and salinity content as means to differentiate depth and length of flow (*ie.*, local, intermediate or regional flow systems).

The position of basement rock becomes an important issue not only for an adequate analysis and understanding of the likely groundwater flow systems that may be governing the hydrogeology of an area, but also for pumping-test results interpretation, for a proper hydraulic potential computation to be used in flow modelling, and of course for the expected water chemistry to be obtained by boreholes; all of which become more important in thick aquifer units.

3 REFERENCE TO STUDY BASINS

Relevant data from three well documented study cases were used to develop evidence on the importance of the effects related to the presence of a thick aquifer unit (*ie* >1,500 m) in the interpretation of observed hydrogeological response.

The examples used are located (figure 1) in different surface topographic basins in three physiographic provinces of Mexico where fractured volcanic rocks prevail. *The Mexico City basin* is situated in the centre of the Mexican Trans-volcanic Belt ($\approx$300 km wide and $\approx$900 km in length) in which major outcrops mainly belong to mafic extrusive rocks. The second *Aguascalientes basin* is located in the Sierra Madre Occidental ($\approx$400 km wide and $\approx$1,600 km in length) where felsic lava flow materials prevail. The third one, the *San Luis Potosi basin*, is located on the eastern boundary of the Central Aluvial Basins ($\approx$200 km wide and $\approx$500 km in length) and is hydraulically bounded by impervious rocks of the Lower Ranges of the Sierra Madre Oriental.

The Mexico City basin incorporates a geological framework in the vertical section, as presented in figure 3. Subsurface lithological features were obtained from direct drilling results and seismic data (Mooser and Molina, 1993) suggesting an aquifer in Tertiary to Quaternary volcanic units (basalt, andesite and rhyolite) as well as related sediments and pyroclastics with a collective thickness in excess of 2,000 m. These volcanic units are extensively fractured and crossed by fault (graben) systems that resulted from tensional regional forces; they partially cover an excess thickness of 1,000 m of limestone strata Cretaceous in age, drilling through these strata reported karstic features (SHCP, 1969). The limestone is laying on an undifferentiated basement rock. Quaternary to Recent deposits form a 30 m to >200 thick aquitard unit (Ortega *et al.*, 1993; Ortega-Guerrero *et al.*, 1999; Edmunds *et al.*, 2002) that outcrops extensively on the *plain*; this remaining lake feature almost fully covers the plain and is highly compressible yielding stored water when the hydraulic potential in the geological material beneath is reduced as a result of active withdrawal boreholes.

The original hydrological setting of this surface basin, as observed before the Conquest in the XVI Century, was that of lakes and the presence of discharge conditions of local, intermediate and regional flows (Ortega and Farvolden, 1989). Some local discharge features still remain in the surrounding mountains. However, steady state conditions evident by the continuous discharge of local, intermediate and regional flows have been disturbed; an example of the severe stress on the discharge areas of local and intermediate systems is the disappearance of the natural conditions of the Xochimilco wetland located in the southern part of the basin, which now survives by an artificial inflow of secondary treated sewage

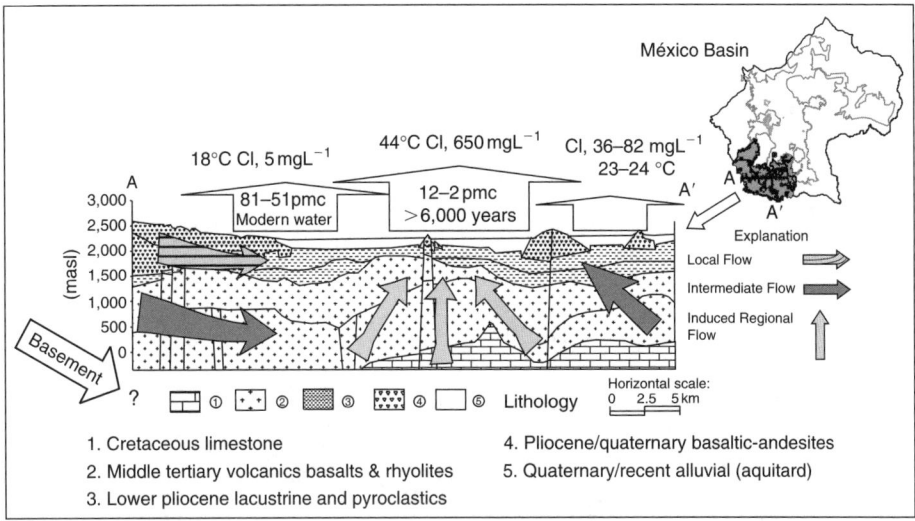

Figure 3. Hydrogeological functioning in the Mexico basin indicating a lithological section in the south of the basin and response of groundwater age from samples collected in boreholes along expected horizontal flow path.

water. Initial pumping-test results in 1970's in nearby boreholes indicated the expected semi-confined conditions resulting from an inflow of the aquitard unit above, overall borehole withdrawal in the basin at that time was claimed to be ≈27 m^3/s. Present water withdrawal to supply the City water needs is estimated in ≈70 m^3/s; >50 m^3/s are groundwater obtained through 100–400 m deep boreholes located on the plain and on the piedmont area. Boreholes located in the piedmont area provide with recent groundwater (intermediate flow) whereas on the plain water age older than 6,000 years is obtained (figure 3). The heavy withdrawal, induces different proportions of regional water from beneath (old water) into shallow withdrawal level; this regional (thermal) inflow and its effect in the pumping-test are usually ignored and often considered as a leaky-effect from the aquitard above. Classical semi-confining conditions have continuously evolved under the heavy withdrawal, shifting this condition to an additional inflow from beneath (Edmunds et al., 2002). In fact, drawdwon vs tme (s-t) curves in each case are basically similar in shape, but not necessarily represent the hydraulic potential evolution with pumping, or the actual drawdown (Huizar et al., 2005).

The Aguascalientes basin is included in a geological framework as that presented in the east-west vertical section of figure 4. Subsurface lithological features were obtained from stratigraphic studies, direct drilling results and electrical resistivity surveys that suggest an aquifer thickness in excess of 1,500 m. The Cainozoic sequence is bounded by Mesozoic sedimentary (mainly calcareous) with the presence of a post-Mesozoic granite intrusive rock. Major rock units are Tertiary volcanic, and undifferentiated granular sediments and pyroclastics Tertiary to Recent in age. The volcanic units are extensively fractured and crossed by fault (graben) systems resulting from tensional regional forces (Hergt, 1997) these volcanic units outcrop extensively on the plain.

The original hydrological setting around the Aguascalientes city was that of a discharge area of regional flow conspicuous by its thermal nature (note, *Aguascalientes* means in Spanish, hot-waters). The extensive development in the basin, and elsewhere along the

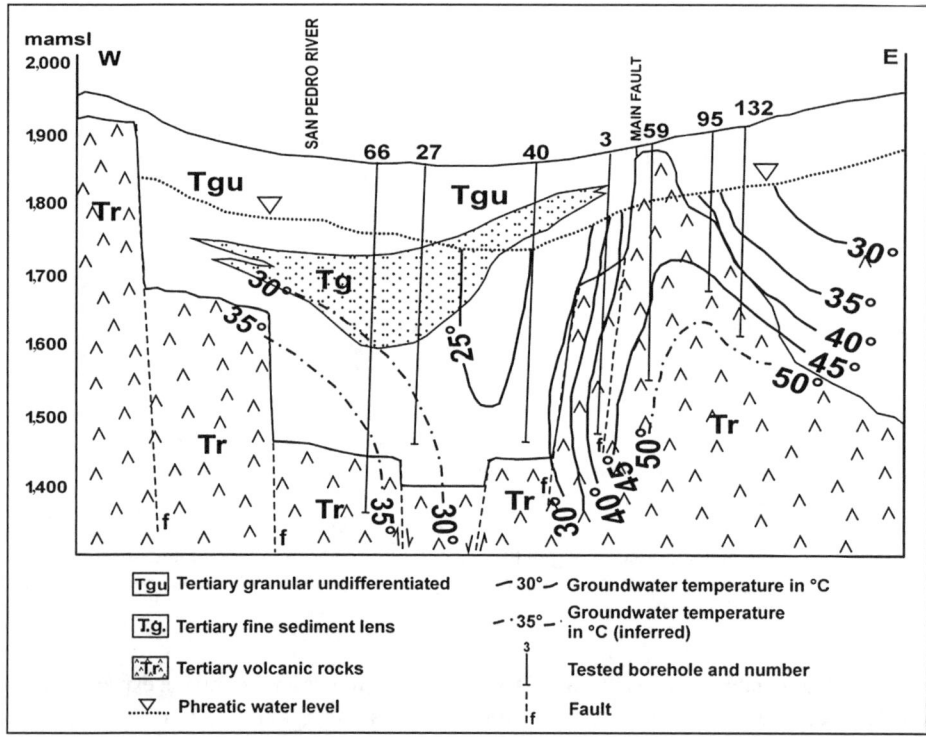

Figure 4. Hydrogeological functioning in the Aguascalientes basin in an east-west section. (Jm) t Mesozoic Basement, (Tr) Tertiary volcanic units, (Tt) Tertiary tuff and clayey sandy sediments, (Tgi) Tertiary granular undifferenciated.

flow path, lowered the potentiometric surface to a depth of more than 70 m; this resulted in the vanishing during the second part of the XX Century, of hot springs, the baseflow in the San Pedro river and the disappearance of related ecosystems. Some local discharge features still remain in the surrounding mountains. Initial pumping-test result in boreholes during the 1970's indicated semi-confined conditions; so, the presence of a semi-confining bed was advocated. By 1996 withdrawal in the central part of the basin was ≈17 m^3/s, groundwater is obtained through 100–550 m deep boreholes located on the plain. This heavy withdrawal has induced thermal water, in different proportions from beneath, into shallow withdrawal level; this thermal inflow, and its effect in the hydraulic response, is usually ignored. In fact s-t curves in each case are basically similar in shape to those representing semi-confined conditions, but not necessarily represent the hydraulic potential evolution with withdrawal time or the actual drawdown (Carrillo-Rivera et al., 2001).

The San Luis Potosi basin incorporates a hydrogeological functioning as shown in the east-west vertical section presented in figure 5. Subsurface lithological features were obtained from direct drilling results, electrical prospecting and magnetic susceptibility data suggesting an aquifer thickness in excess of 1,500 m. The aquifer unit consists of Tertiary volcanic units of felsic composition (ignimbrites, lava flows and tuffs) as well as related sediments derived from limestone and volcanic rocks and pyroclastics (collectively named Tertiary Granular Undifferenciated). These volcanic rocks have been extensively

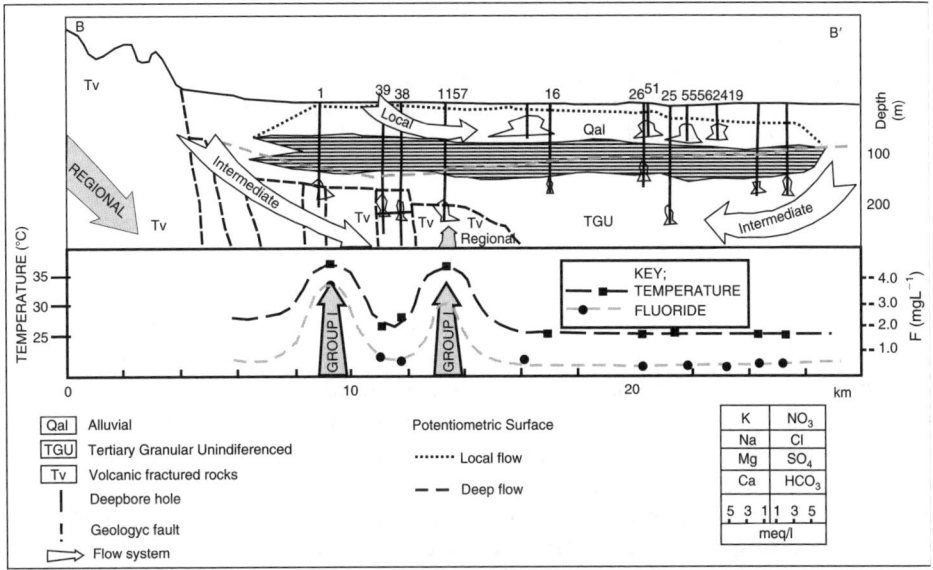

Figure 5. Hydrogeological functioning in the San Luis Potosi basin in an east-west section. Groundwater temperature response in pumping boreholes is depicted along the expected horizontal flow path direction.

fractured and crossed by fault (graben) systems that result from tensional regional forces. The volcanic sequence partially covers Mesozoic sedimentary (mainly calcareous) units with the presence of a post-Mesozoic granite intrusive. Some 100 m below surface level a low permeability compact fine sand lens ($K = 10^{-9}$ m/s) subcrops the plain except its margins, its maximum thickness is about 50 m. Available data (Carrillo-Rivera et al., 1996) suggests the presence of two flow systems: one is represented by thermal water (40.4°C) related to a regional flow system, as suggested by B (0.17 mgL^{-1}), F$^-$ (3.1 mgL^{-1}), Na$^+$ (53.2 mgL^{-1}), and Li$^+$ (0.19 mgL^{-1}) concentrations that imply a large residence time and interaction with rhyolitic rocks. The second flow is represented by cold water (25.5°C ± 1°C) that implies an intermediate flow system with low content of B (0.03 mgL^{-1}), F$^-$ (0.4 mgL^{-1}), Na$^+$ (14.6 mgL^{-1}), Li$^+$ (0.01 mgL^{-1}) which suggests short residence time and interaction with the granular material. The original hydrological setting of this surface basin was that of a transit area of regional and intermediate flow systems (the first deep borehole drilled in 1940 was 160 m deep and had a water level depth of 100 m), so a lack of discharge conditions were present advocating the presence of interbasinal flow (Carrillo-Rivera, 2000). The only identified discharge features in the basin, are those of local flow systems in the surrounding mountains. Initial pumping-test results in 1970's in boreholes obtaining thermal water indicated semi-confined conditions resulting from an inflow of regional flow from beneath, withdrawal in the basin during 1972 was claimed to be ≈0.8 m^3/s with an "average drawdown" of 0.9 m/year; in 1977 the reported withdrawal was of 1.9 m^3/s which produced an "average drawdown" of 1.0 m/year; however, in 1987 the reported withdrawal of 2.6 m^3/s produced an "average drawdown" of only 1.3 m/year (Carrillo-Rivera, 2000). Present water withdrawal to supply the City water needs is estimated in about 4 m^3/s obtained mainly through 300–500 m deep boreholes located on the plain. Withdrawal has induced groundwater of intermediate and thermal flow systems in different proportions; the thermal inflow from beneath affects potentiometric measurements and pumping-test results. The s-t pumping-test curves are basically similar in shape to those of semi-confining conditions, but field data fails to represent the hydraulic potential evolution with withdrawal or the actual drawdown.

4. CHANGES IN CHEMISTRY AND TEMPERATURE FOR VERTICAL FLOW IDENTIFICATION DURING GROUNDWATER EXTRACTION

Actual groundwater movement in the vicinity of a borehole site may have important vertical components of flow due to heterogeneity of aquifer units, the depth of the borehole, density of the borehole-field, and magnitude of groundwater extraction in nearby boreholes, as well as the nature of the flow systems present in aquifer units. The vertical components of flow in a pumping borehole may not be satisfactorily recognized from horizontal hydraulic surface representation, where the position of flow lines could be affected by lithology contrast, difference between depth of borehole screen location and flow system conditions in terms of recharge or discharge areas. These effects wait to be included in the methodology of analytical pumping-test data interpretation. However, the presence and nature of the vertical components of flow developed during a pumping-test may be estimated by understanding the flow regime in the tested borehole.

Usually, pumping-test requires relevant data to be collected, mainly withdrawal yield and variation of depth to water level in time since the test started. Measurements of total dissolved solids (TDS), temperature, electrical conductivity (EC) as well as pH, Eh, dissolved oxygen obtained in the discharge water during the test could be useful in identifying flow components other than from horizontal flow. Figure 6-A shows flow response in a borehole due to withdrawal when horizontal flow prevails in a confined aquifer, log-log representation of s-t data will follow the shape of the *Theis curve* (Theis, 1935); the expected variation of EC (or TDS) and temperature with withdrawal time, in obtained water, is negligible. However, in a thick aquifer the log-log s-t data fails to respond as that from a confined aquifer unit and shows a "recovery", prevailing field flow conditions need to be reviewed. The log-log s-t data will achieve a similar shape as that described for semi-confined conditions (Hantush, 1956) when leakage from an overlaying aquitard influences the test (figures 6-B, 6-C); TDS content may be reduced or incresed (curve "a" or "b") according with the salinity contrast between water in the aquitard and that from the aquifer unit beneath. Just as water temperature is to be reduced when the inflow of shallow water from an overlaynig aquitard is expected to influence the test, the outflow temperature will increase when a deep flow system has been induced to the shallow production level of the borehole (figure 6-C).

A possible procedure to deduce the control of the incoming flow may be reached by interpreting TDS (or EC) and temperature data measured contemporaneous and continuously along a pumping-test; an additional understanding of the flow to the borehole could be achieved by using pH and chemical data such as Eh, DO or specific ions (Carrillo-Rivera et al., 1996, Huizar-Alvarez et al., 2004). The chemical and physical response of groundwater withdrawal is considered to provide information on the importance of the hydraulic characteristics of the geological media influencing the test and the nature of the flow thus developed. These information is required to make considerations accordingly (*ie*, hydraulic head corrections by density difference effect) as from where other hydrogeological aspects could be proposed (*ie*, preferential flow from a geological unit or from a particular

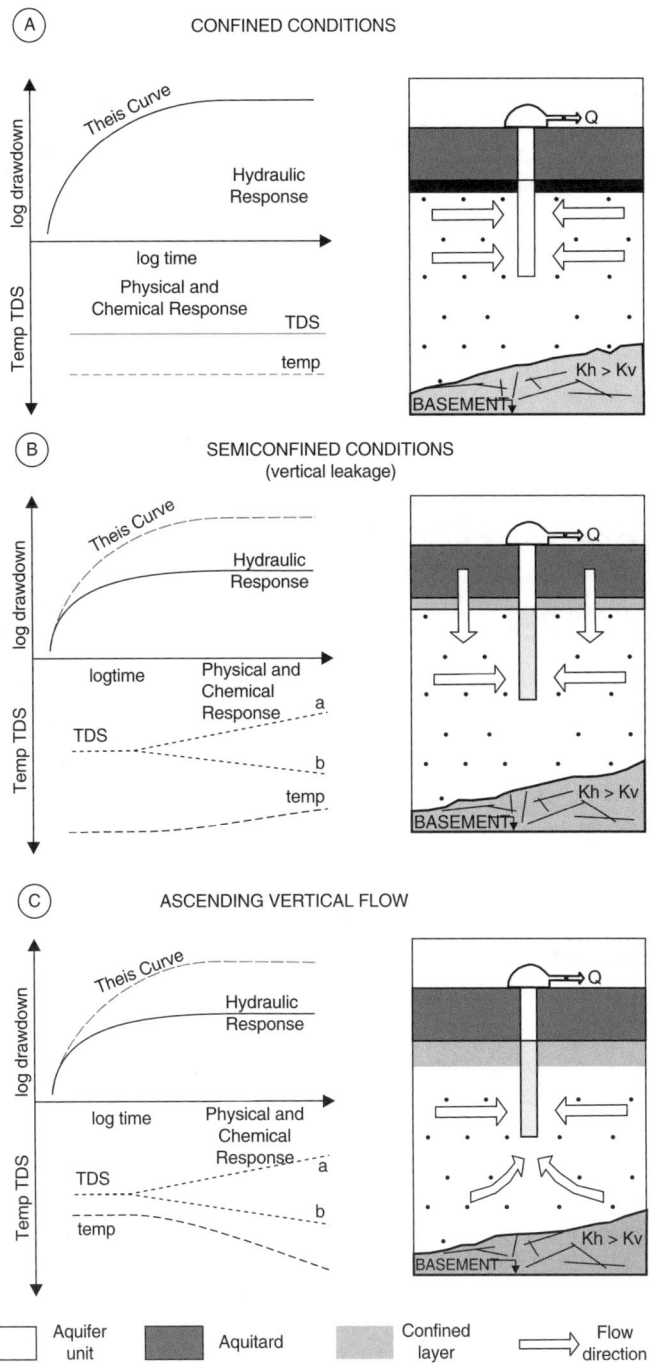

Figure 6. Theoretical expected response in terms of *s-t* as well as temperature and TDS of extracted grundwater for: (A) a confined aquifer with horizontal flow to the pumped borehole, (B) with vertical flow derived from leaky effect resulting from semi-confined conditions, and (C) vertical inflow of water from beneath the borehole pumping level.

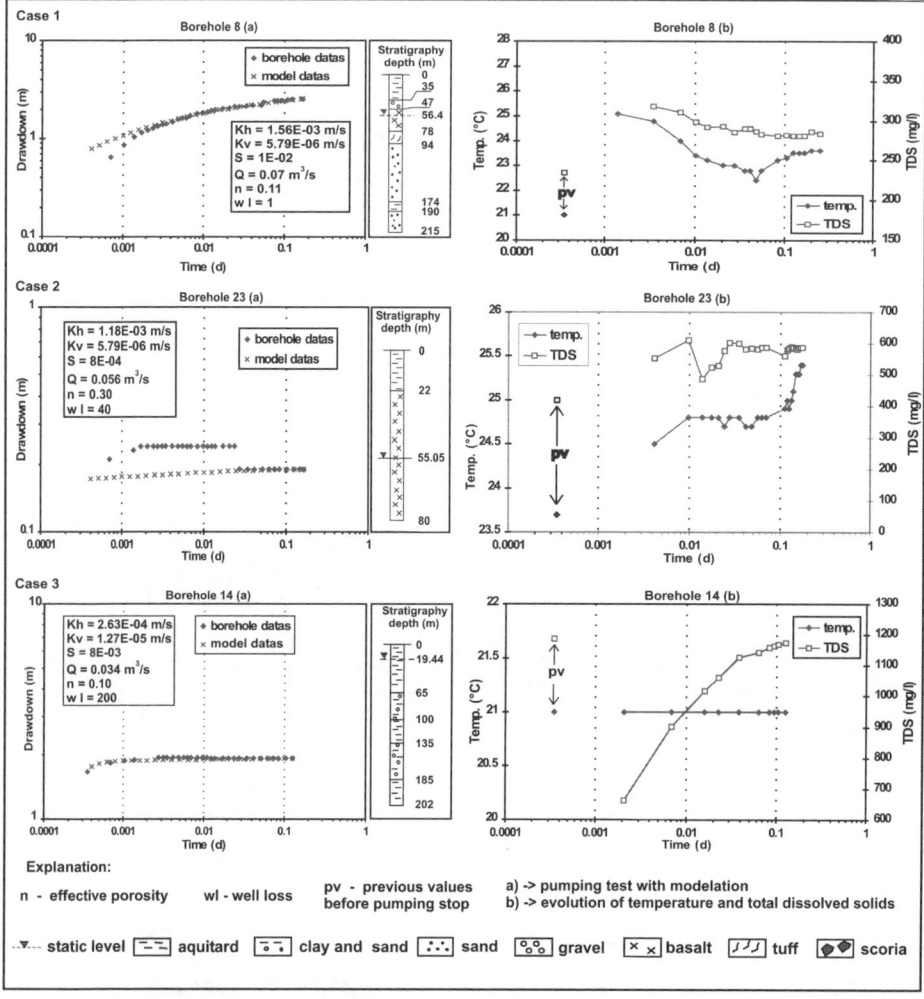

Figure 7. Rate of drawdown with time for Mexico City boreholes 8, 23 and 12 where the development of temperature and TDS with pumping time assist to define confined conditions (borehole 8 and 12) and upward flow (borehole 23). After Huizar et al., 2004.

flow system). An adequate interpretation of water temperature and TDS (or chemistry) could prove a useful tool to identify the nature of the vertical components of flow in a pumping-test as well as the nature of the of the scale of the flow that is obtained in the borehole.

Three real cases of the Mexico City basin are examined. Figure 7, case 1 (borehole 8) proposes a leaky effect from the semi-confinig unit (above) that results in the inflow of cold water; lond term withdrawal would produce water with lower temperature than along the test, to about 21°C and a change in the chemical characteristics (reduction of TDS) of the obtained water. Drawdown-time data in figure 7, case 2 (borehole 23) would suggest a standard leaky response where the drawdown rate is diminished along the duration of the test; however, the temperature increase of the inflow water may suggest that water is induced from beneath. The response of the increase in TDS and constant temperature of case 3 in figure 7 suggests that obtained water is derived from standard confined conditions. This may imply that the inflow to the pumping borehole is controlled by the hydrogeological characteristics of the aquifer media, where the effect of the drawdown on the hydraulic potential, the length of the path to travel by any flow system, and the ratio between K_v/K_h are to be understood, so the prevailing flow mechanisms may provide with a possibility control of the type of flow that may be induced into the borehole.

5 CONTROL OF VERTICAL FLOW TO PREVENT WATER QUALITY IMPAIRMENT

The identification of the different flow systems entering a pumping borehole is considered an available tool for possible control of obtained water quality (Carrillo-Rivera *et al.*, 2002). Figure 8 shows borehole 61 located in granular material on the plain of San Luis Potosi basin producing water with similar chemical composition along the 20 hours of the duration of the test (no pumping-test available). However, borehole 34, with similar depth to borehole 61, located in fractured rock suggests that during the 60 hours of the test the inflow of water develops a contrasting chemical behavior from a cold and low fluoride and TDS content, to a thermal and high fluoride and TDS concentration.

Other factors affecting groundwater flow distribution in the vicinity of a pumping borehole could depend on its construction, lithology composition, flow systems affected and their contrast in chemical composition. In the San Luis Potosí basin, the tapped intermediate flow system with cold water, and the regional system with thermal water, also have contrasting chemistry where fluoride content might be used as an indicator of the prevailing flow conditions. A borehole might withdraw water according to the following five scenarios as shown in figure 9. Case (A) suggests a *3D* radial flow to a borehole with a mixture in different proportions of the two flow systems. Case (B) implies the regional (thermal) water from beneath enters the borehole after groundwater passes through Tertiary Granular Undifferenciated (TGU) where its flow velocity will be reduced and some reactions might occur in the TGU and changes in the physical (reduction in temperature) and chemical content (precipitation of fluoride in the TGU) are expected (Carrillo-Rivera *et al.*, 2002). Case (C) proposes that intermediate (cold) flow is obtained with higher fluoride content than that expected from this flow system; chemical composition that might be explained acknowledging that intermediate water flowing through TGU is induced into the fractured media where it obtains its fluoride content. In case (D) regional flow water with high fluoride is obtained directly from the fractured material. In case (E) only cold intermediate water with low flouride is reaching the borehole through the TGU. Pumping-test related to cases (C) and (E) are expected to exhibit *s-t* results showing the prevailing confining conditions, and constant low water temperature indicating the presence of horizontal groundwater flow conditions.

6 REGIONAL DISTRIBUTION IN SHALLOW AQUIFER UNITS OF INDUCED WATER BY DEEP BOREHOLE WITHDRAWAL

Groundwater in shallow aquifer units creates concerns on the possibility of becoming contaminated by different antropic activities. Usually, expected contaminants may reach the flow systems from the soil surface. However, the importance of processes that deteriorate obtained water quality by excessive withdrawal through the inducement of flows from

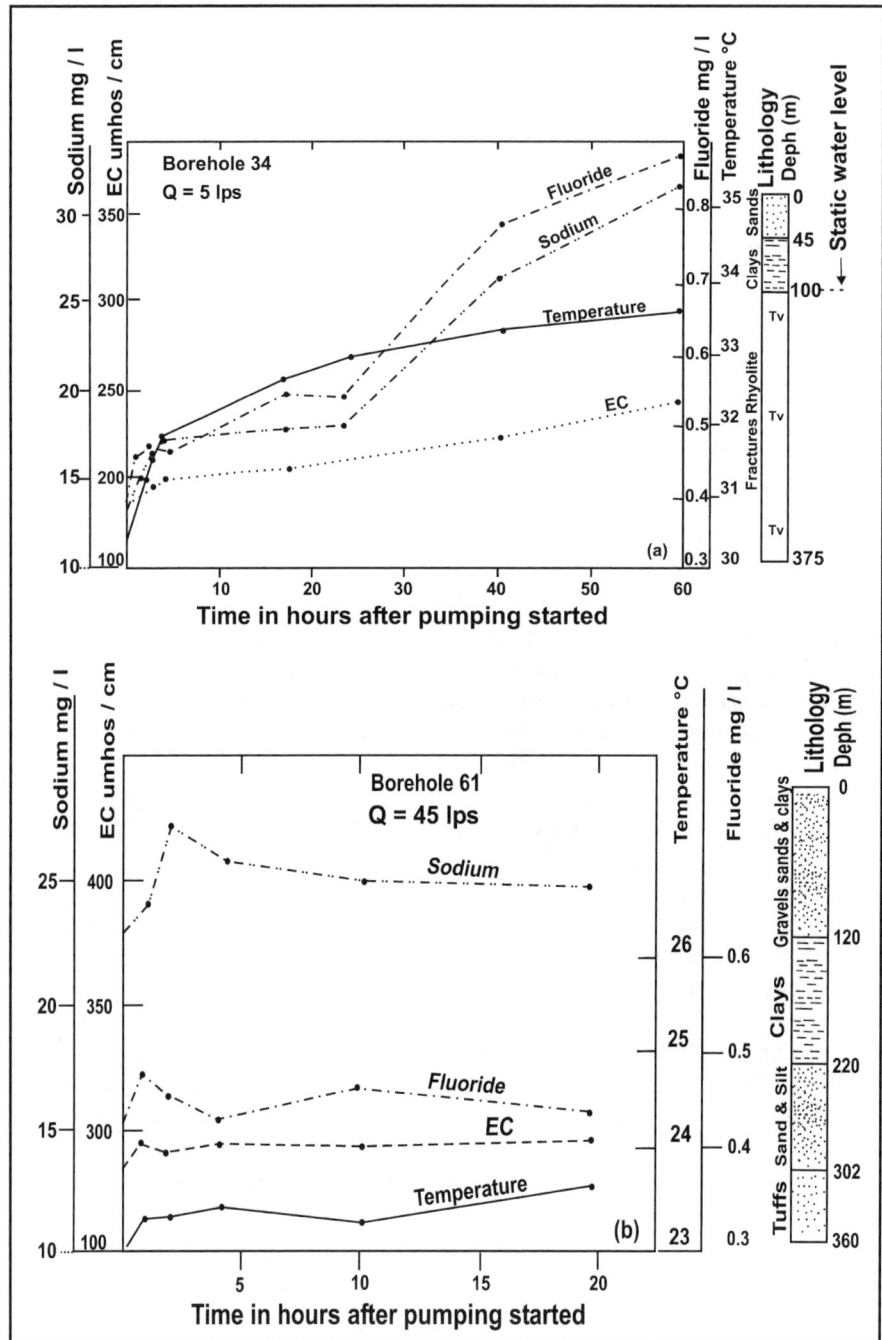

Figure 8. Water quality response with pumping time in boreholes with vertical induced input from beneath (borehole 34) and horizontal flow (borehole 61) adapted from Carrillo-Rivera et al., 1995.

Groundwater flow system response in thick aquifer units: theory and practice in Mexico 41

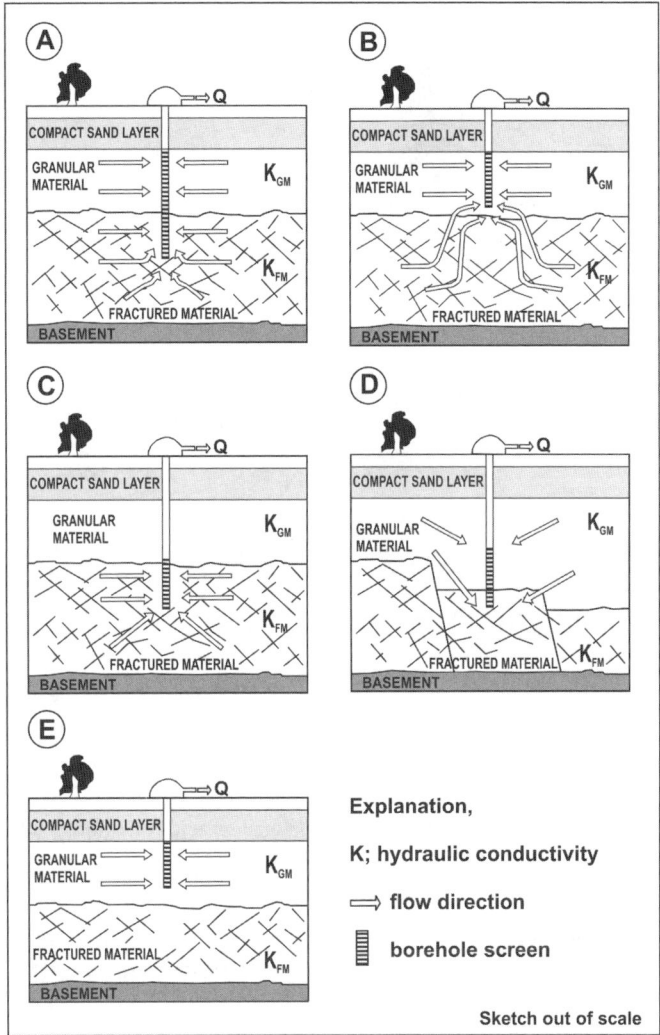

Figure 9. Groundwater flow circulation in the vicinity of a borehole depending on their design and overall lithology. TGU, means Tertiary Granular Undifferenciated construction (adapted from Carrillo-Rivera *et al.*, 2002)

beneath with undesirable chemical constituents has lacked the required attention. This impact is more likely to occur in aquifers of great thickness in which flows of regional hierarchy with undesirable chemical constituents are present. In México, the first constructed boreholes obtained good quality water from shallow levels. Based on this response, recent enterprises in industrial agriculture enhance district developments creating an accumulative response by the consequent additional withdrawal that induces high salinity and sodium regional flow, resulting in ithe increasing alkalinity and pH of the soil. Initial soil salinity and sodium deposition effects were tackled by an additional groundwater extraction that in turn increases the inducement of regional flow. Figure 10 shows induced upward vertical movement of regional groundwater flow due to the drawdown produced in a borehole

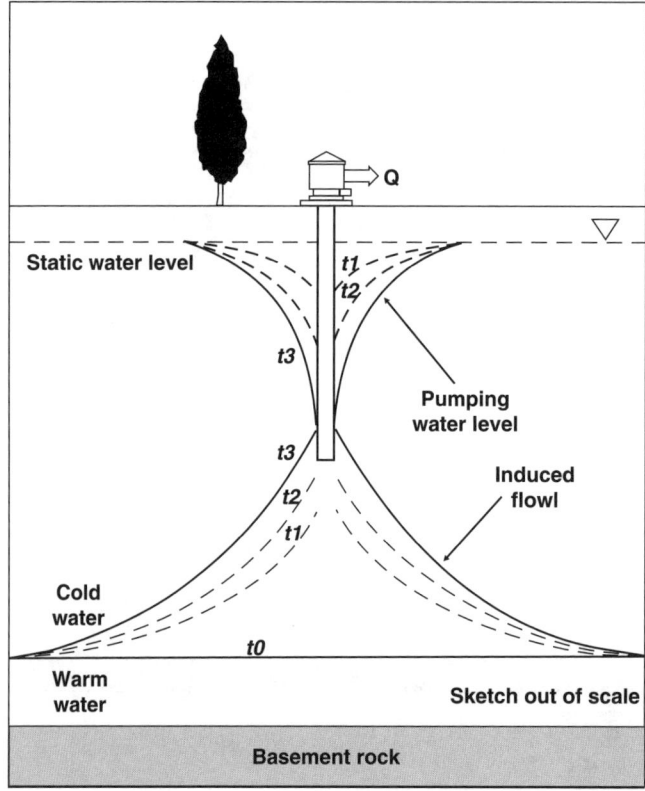

Figure 10. Induced upward vertical movement of regional groundwater flow due to the drawdown produced in a borehole partially penetrating a thick aquifer unit.

partially penetrating a thick aquifer unit; the *density driven flow* of the regional water front is proportional to the reduction of the cold water thickness, the vertical hydraulic conductivity and the density difference between the cold and the thermal flow. This kind of developed flow is termed at times as *buoyancy induced flow* (Holzbecher, 1998). The rise of regional water to shallow levels is not a classical contamination point source problem; measurements made (figure 11a) intent to define the extent of the inducement of the regional system from the San Ignacio borehole in the Aguascalientes basin. Temperature was used as an indicator of the rise of the regional flow beyond the screen interval of a withdrawal borehole. An assessment of the radius of influence of the effect of temperature rise due to withdrawal still waits to be defined. This effect of regional groundwater inducement to shallow levels is translated to an overall increase in the temperature in the obtained groundwater and is also reflected as an increase in other undesirable dissolved elements (Carrillo-Rivera *et al.*, 2002). Figure 11b presents observed evolution of the depth to the dynamic water level during a step drawdown test, suggesting a natural adjustment to the hydraulic potential in the water level elevation due to the inflow of water with high temperature.

In this regard groundwater obtained in the Aguascalientes basin (as many other surface basins in central Mexico) has experienced a temperature increase; figure 12 shows a

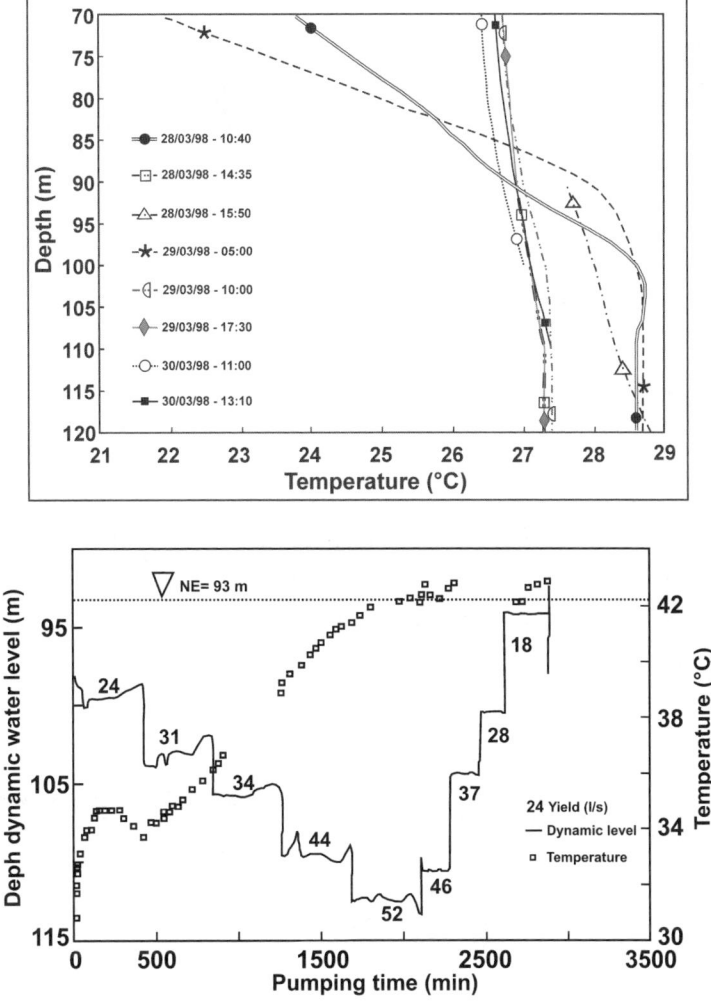

Figure 11. (a) Temperature distribution at depth in an observation borehole for time intervals as shown, measurements were collected during a step-drawdown test carried out in the San Ignacio borehole located some 30 m distance. (b) Withdrawal yield, depth to dynamic water level and temperature results in tested San Ignacio borehole (adapted from Carrillo-Rivera et al., 2001).

difference of available temperature data from 1971 to 1995 indicating a rise of more than 10°C. This implies a regional implication in hydraulic related computations as well as in water quality for agriculture and domestic supply purposes as the outflow is rich in sodium content.

7 CONCLUSIONS

The final remarks about the development of groundwater from aquifer units that have a large thickness ($\geq 1{,}500$ m) may be incorporated in two categories: one, methodological considerations on groundwater hydraulic computations; and a second, the practical implications in

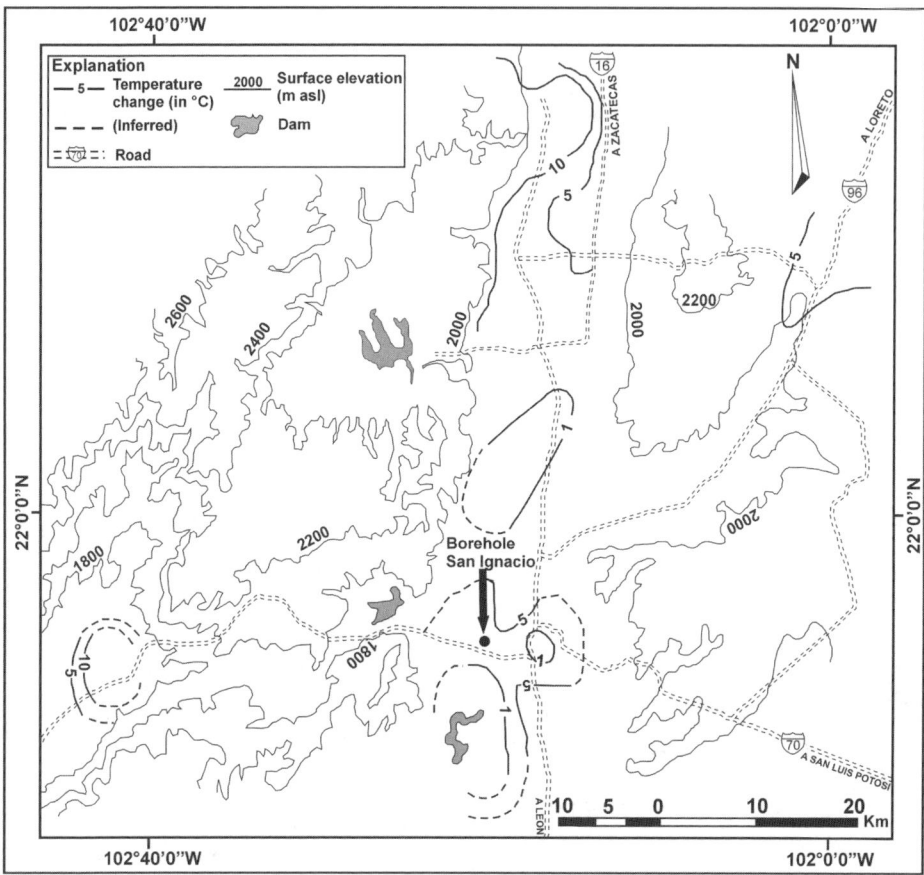

Figure 12. Difference in temperature from 1971 to 1995 measured at discharge-head in boreholes located in the Aguascalientes basin. Location of San Ignacio borehole.

the short and long scope regarding the impact on the quality of obtained water. A final link may be argued by Carrillo-Rivera (2000) where the regional extensión and thickness of aquifer units permit to assume the presence of large flow paths that are responsible of interbasinal flow; such flows, and related connectionmust be considered when groundwater is to be evaluated in a given surface hydrologic catchment.

An understanding of the type of groundwater functioning developed in a thick aquifer unit may be incorporated in any methodology through adecuate groundwater flow modelling that includes boundaries such as depth to basement rock and response in terms of variable density flow. Such evaluation includes the need to incorporate the scales of groundwater flow within the prevailing geological media.

Pumping-test analysis and groundwater balances that are currently carried out may prove a valuable tool when integrated in the natural media of the sites where they are applied into, and whose hydraulic control is represented by the flow systems. The quality of the water that is currently obtained by boreholes is changing with time suggesting a different inflow to that resulting from purely horizontal flow derived from capturing local or intermediate systems.

This suggests that long term evaluation and availability of groundwater requires, on one hand, to incorporate groundwater flow that is derived beyond the area of computation, and on the other hand, the effects in areas hundreds of kilometers away that are connected to the computation area, where possible effects on ecosystems might be taking place.

Groundwater flow in thick aquifer units requires further understanding and consideration to explain water quality changes by ascending thermal flows, whose impact in time and space has become an important issues in groundwater flow understanding, as a means to achieve its sustainable management from the local to a regional scale.

REFERENCES

Bear J (1972) Dynamics of Fluids in Porous Materials, American Elsevier Publishing Company, New York, NY.
Boulton NS (1954) Unsteady radial flow to a pumped well allowing for delayed yield from storage. Proccidings, International Asociation of Scintific Hydrology, Pub 37: 472–477
Boulton NS and Streltsova TDS (1978) Unsteady flow to a pumped well in an unconfined fissured aquifer. Journal of Hydrology 37: 349–363
Carrillo-Rivera JJ, Cardona A and Moss D (1996) Importance of the vertical component of groundwater flow: a hydrogeochemical approach in the valley of San Luis Potosi, Mexico. Journal of Hydrology 185: 23–44
Carrillo-Rivera JJ (2000) Application of the groundwater-balance equation to indicate interbasin and vertical flow in two semi-arid drainage basins, México. Hydrogeology Journal 8(5): 503–520
Carrillo-Rivera JJ, Cardona A and Hergt T (2001) Inducción de agua termal profunda a zonas someras: Aguascalientes, México. Revista Latinoamericana de Hidrología 1(1): 41–53
Carrillo-Rivera JJ, Cardona A and Edmunds WM (2002) Use of abstraction regime and knowledge of hydrogeological conditions to control high fluoride concentration in abstracted groundwater: basin of San Luis Potosi, Mexico. Journal of Hydrology 261: 24–47
Edmunds WM, Carrillo-Rivera JJ and Cardona A (2002) Geochemical evolution of groundwater beneath Mexico city. Journal of Hydrology 258: 1–24
Hantush MS (1956) Analysis of data from pumping test in leaky aquifers. Transactions of the American Geophysical Union. 37(6): 702–714
Hergt T (1997) Untersuchugen der vertikalen hydraulischen Leitfähigkeit in gestörten Festgesteinsserien bei Aguascalientes, Mexico. Diploma Thesis, Institute fur Angewandte Geowissenschaften II, Berlin
Hergt T and Carrillo-Rivera JJ (2004) Investigación de la conductividad hidráulica vertical en un medio de gran espesor. Proceedings, XXXIII AIH and 7° ALHSUD Joint Congress, Zacatecas, México
Holzbecher E (1998) Modeling density-driven Flow in porous media. Springer Verlag, Berlin
Hubbert KM (1940) The theory of groundwater motion. Journal of Geology 48(8): 785–944
Huizar-Alvarez R, Carrillo-Rivera JJ, Angeles–Serrano G, Hergt T and Cardona A (2004) Chemical response to groundwater withdrawal southeast of México City. Hydrogeology Journal 12(4): 436–450
Kawecki MW (1995) Correction for Temperature Effect in the Recovery of a Pumped Well. Ground Water 33(6): 917–926
McDonald MG and Harbough BR (1988) A modular three-dimensional finite difference ground water flow model. USGS Techniques of water resources investigation report, Book 6, chapter A1
Mooser F and Molina C (1993) Nuevo modelo hidrogeológico para la Cuenca de Mexico. Boletín del Centro de Investigación Sísmica de la Fundación Javier Barros Sierra 3(1): 68–84
Ortega A and Farvolden RN (1989) Computer analysis of regional groundwater flow and boundary conditions in the basin of Mexico. Journal of Hydrology 110: 271–294
Ortega A, Cherry JA and Rudolph DL (1993) Large-scale aquitard consolidation near Mexico City. Groundwater 31(5): 708–718

Ortega-Guerrero, MA, Rudolph, LD and Cherry, JA (1999) Analysis of long-term land subsidence near Mexico City: field investigations and predictive modelling. Water Resources Research 35(11) 3327–3341

Rathod KS and Rushton KR (1991) Interpretation of pumping from two-zone layered aquifers using a numerical model. Ground Water 29(4) 499–509

SHCP (Secretaría de Hacienda y Crédito Público) (1969) El hundimiento de la ciudad de México, Proyecto Texcoco [The sinking of Mexico City, Texcoco Project], Volumen Nabor Carrillo, México

Theis CV (1935) Relation between the lowering of the piezometric suface and the rate and duration of discharge of a well using groundwater storage. Transactions of the American Geophysical Union 16: 519–614

Tóth J (1962) Theory of groundwater motion in small drainage basins in central Alberta, Canada. Journal of Geophysical Research 67(11): 4375–4387

Tóth J (1963) Theoretical analysis of groundwater in small drainage basins. Journal of Geophysical Research 68: 4791–4812

Tóth J (1995) Hydraulic continuity in large sedimentary basins. Hydrogeology Journal 3 (4–16)

Tóth J (1999) Groundwater as a geologic agent: an overview of the causes, processes, and manifestations, Hydrogeology Journal 7(1): 1–14

CHAPTER 1

Integrative modelling of global change effects on the water cycle in the upper Danube catchment (Germany) – the groundwater management perspective

Roland Barthel[1], Wolfram Mauser[2] and Juergen Braun[1]
[1]Institute of Hydraulic Engineering (IWS), Universitaet Stuttgart, Stuttgart, Germany
[2]Dept. of Earth and Environmental Sciences, Chair for Geography and Remote Sensing,
Ludwig-Maximilians University (LMU) München, München, Germany,
Corresponding author: **Roland Barthel**[1]

ABSTRACT: The GLOWA program addresses the manifold consequences of Global Change on regional water resources in a variety of medium sized watersheds. The Upper Danube Basin (Germany, Austria; of about 77,000 km^2) represents a mountain foreland situation in the temperate mid-latitudes. The major goal of "GLOWA-Danube" is the development of new water resource management modelling technologies, integrating both natural and socio-economic sciences. GLOWA-Danube relies on the Decision Support System (DSS) DANUBIA, a coupled system that is based on currently 16 individual sub-models which are developed by 11 different research groups from different disciplines at different locations (mainly German Universities). Each model runs on a different computer and exchanges data with partner models using the common DANUBIA architecture and Internet protocols. DANUBIA and its sub-components are object-oriented, spatially distributed and raster-based and have been developed using the Unified Modelling Language (UML) and JAVA. This paper describes the framework of GLOWA-Danube and the integrated model DANUBIA with a focus on two models developed by the Universitaet Stuttgart. These two sub-models "Groundwater" and "WaterSupply" form together the Groundwater Management complex of DANUBIA which are described in this paper.

Keywords: Global Change, Integrated Water Resources Management (IWRM), Groundwater Model, Danube, Germany.

1 INTRODUCTION

Recently, integrated approaches for describing, modelling, and forecasting physical, social, economic, and political processes related to the hydrological cycle, in particular with regards to Global Change, have gained worldwide attention both with administrative authorities and in the research community. Water affects all economic, cultural, social and ecological aspects of daily life. It is the basis of functioning matter cycles and hence of a clean, stable and sustainable environment. A functional understanding of the processes related to the water cycle and the influence of human societies upon these is crucial for the development of ways for the sustainable management of water. Since the related processes are strongly inter-related, sectoral science approaches are neither capable of understanding the complex interactions between nature, water and man nor of developing methods for a sustainable water resource management under globally changing boundary conditions. A high level of trans-disciplinary integration is required to provide a profound scientific knowledge base, taking into account continuously changing natural, social and technological boundary conditions known as Global Environmental Change. Proactive watershed management aiming at a sustainable use of the water resources thus relies heavily on the development of future scenarios and on numerical models with predictive abilities. To date, no commonly accepted modelling approaches are available to integratively describe the complex interactions between natural and social processes. The lack of successful integration concepts is the result of large differences in the way the various disciplines formalize and describe their understanding of the respective processes. These differences in terms and concepts, comprehension and methodology lead to sectoral approaches for solving separate parts of the task, and hence provide no reliable basis for simulating recursive and interactive scenarios of future development.

Increasing intensity of water use and water-related conflicts between numerous stakeholders puts increasing pressure on the natural environment and ecology. Stakeholders represent governments, society, nature and industry. DSS try to combine both comprehensive modeling with decision-making and stakeholder support. Developing and using a DSS is expected to aid in the following tasks: structuring of problems, integration, information analysis learning and, of course, decision making. In this way a DSS can facilitate discussion between the parties involved in environmental and security management issues. A DSS provides an arena where short- and long-term impacts of proposed actions can be observed (in time and space) and where the feasibility of actions can be investigated. To determine the sustainability of various management alternatives and to derive appropriate recommendations for public and commercial stakeholders it is necessary to accurately describe the complexity of water-related issues by an integrated approach. For decision-making purposes, indicators representing the driving forces of change and simplifying complex information must be identified.

Many examples of Decision Support Systems can be found in the literature and many projects in this regard have been carried out. These approaches usually deal with isolated water-related problems and little effort has gone into making this scientific material available as part of practical planning or management tools for public policy makers at the regional level. The objective of GLOWA-Danube is to provide new modelling technologies to overcome this discrepancy and provide a common basis for scientific analysis and planning practices.

2 THE GLOWA-DANUBE PROJECT

Within the GLOWA-initiative (Global Change of the Water Cycle, www.glowa.org, funded by the German Ministry of Research and Education (BMBF), BMBF, 2002; BMBF, 2005), the Upper Danube watershed (figure 1) was selected as a representative mesoscale test site in the temperate mid-latitudes.

The interdisciplinary research co-operation "GLOWA-Danube" is developing the Global Change DSS "DANUBIA" to investigate the sustainability of future water resources management alternatives. The system equally considers the influence of natural changes in

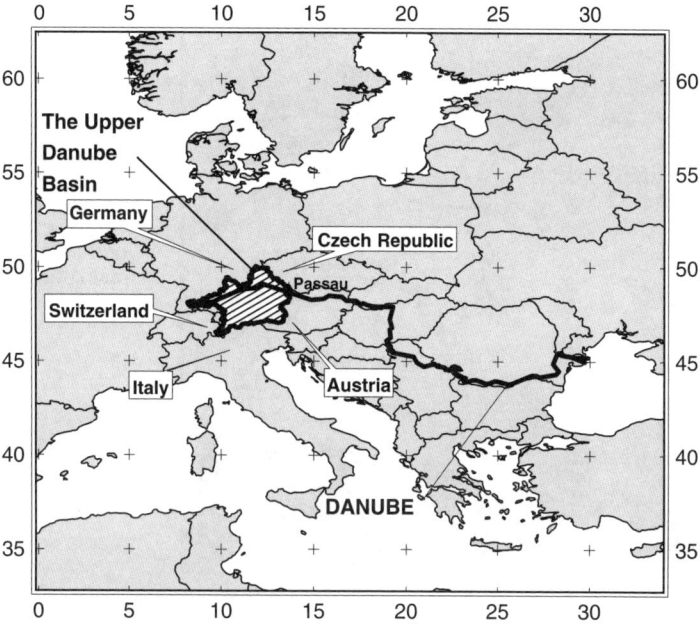

Figure 1. The location of the Upper Danube Basin in central Europe.

the ecosystem, such as climate change, and changes in human behaviour, e.g. changes in land use or water consumption (Mauser and Barthel, 2004; Ludwig et al., 2003). GLOWA-Danube comprises a university-based network of experts combining water-related competence in the fields of engineering, natural and social sciences. The project consists of the following disciplinary research groups which cover the essential modules in GLOWA-Danube: Coordination and GIS, Remote Sensing – Hydrology, Meteorology, Water Resources Management – Groundwater, Water Resources Management – Surface Waters, Plant Ecology, Environmental Psychology, Environmental Economics, Agricultural Economics, Glaciology, Remote Sensing – Meteorology, Tourism Research and Computer Sciences. In the first phase of the project (2001–2004), a prototype of the DSS "DANUBIA" comprising 16 fully coupled disciplinary models was developed and is now in a stage of validation and refinement. The second project phase (2004–2007) started in March 2004. While the focus of the first phase was on the development of technical solutions, process descriptions, definition of exchange parameters and interfaces, and disciplinary model development, in the second phase scenario evaluation, stakeholder involvement, decision making, and water management support are at the centre of the research activities. However, it has become apparent that there is still a need for detailed basic research in several parts of the system to gain a better understanding of the processes involved. The coupling of groundwater, surface water, and land surface models is an example of such shortcomings.

2.1 *The Upper Danube Basin*

With a watershed-area of 8,17,000 km^2 shared by 15 countries, the Danube is the second largest river in Europe (figure 1). GLOWA-Danube is restricted to the analysis of the

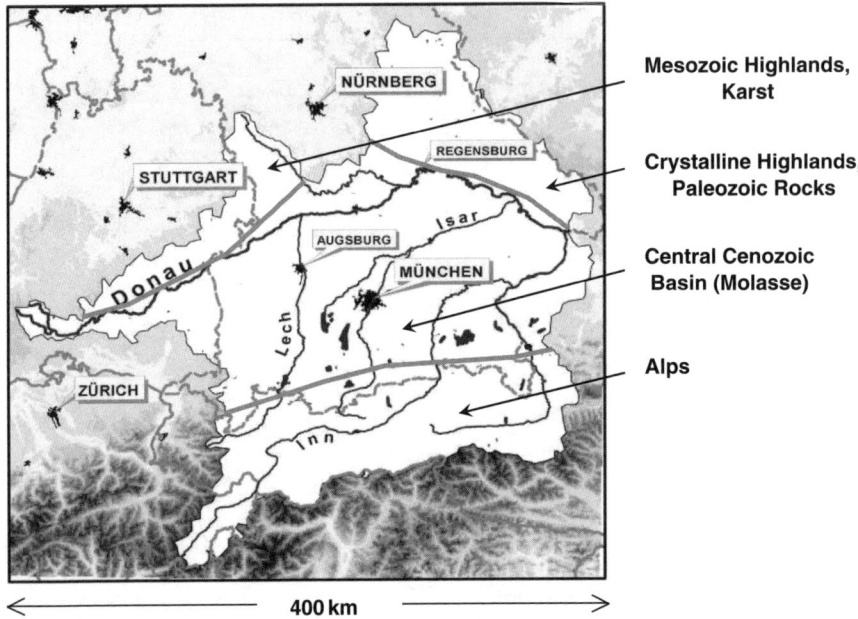

Figure 2. The Upper Danube Catchment.

Upper Danube (A~77,000 km^2), which is defined by the discharge gauge Achleiten near Passau in Germany. The Upper Danube is a mountainous catchment with altitudes ranging from 287 to 4,049 m a.s.l. and a large foreland (figure 2). This introduces strong geographic, meteorological and socio-economic gradients (precipitation: 650 to >2,000 mm/a, evaporation: 450–550 mm/a, discharge: 150–1,600 mm/a, average annual temperature: −4.8 to +9°C, sources of income changing from industry and services to agriculture and tourism). The highly fragmented land cover and land-use is mostly determined by human intervention. Forestry and agricultural use of differing intensity (grassland, farmland) dominate, whereby climatic disadvantages in terms of high precipitation and low temperatures limit the present agricultural potential in various parts of the catchment (Mauser and Barthel, 2004).

Water resource management in the Upper Danube is complex, in part because the area extends over a number of countries: 73% of the Upper Danube is managed by the German states Bavaria and Baden-Württemberg, 24% by Austria and the rest by Switzerland, Italy and the Czech Republic. The Inn (figure 2), as the most important **alpine** tributary, contributes up to 52% of the average discharge of 1,420 m^3/s as a result of very high precipitation and subsequently high surface and subsurface runoff (figure 3).

The Upper Danube is densely populated with approximately 8 million inhabitants. A large part of the water for the water supply of the larger cities and industry originates in the pre-alpine region and in the Alps. The most important industrial agglomeration areas are Munich (1.2 Million inhabitants), Augsburg (2,60,000), Ingolstadt (1,15,000) and the "chemical triangle" Burghausen.

For flood protection, energy production and water-resources-management purposes, the discharge of all important tributaries of the Upper Danube has been regulated through reservoirs and dams. To a large extent, their management is determined by the dynamics of

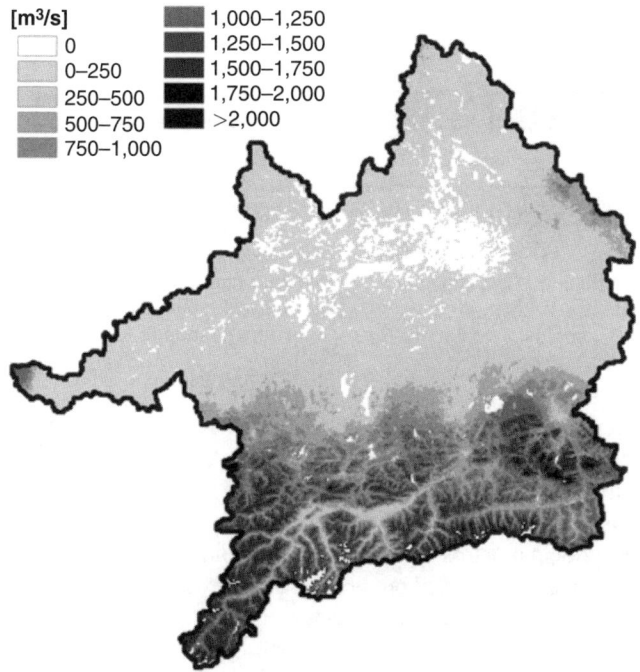

Figure 3. Mean annual groundwater recharge (validation period 1995–1999) calculated by DANUBIA.

the snow and ice storage in the Alps. Reservoir management at present is largely uncoordinated. Therefore a large potential for optimisation of the management practices exists. Parts of the Upper Danube are navigable and are part of an important waterway that connects the Black Sea with the North Sea. This waterway is already used to export water from the catchment area of the Upper Danube into the catchment of the River Rhine. Increasing demand for water during the course of a more intense and more coordinated water use in Europe will put increasing pressure to export more water from the catchment area of the Upper Danube. In general, the ecological and socio-economic effects of water resources use and hence the limits to an environmentally sound water use are still largely unexplored.

2.2 The Decision Support System DANUBIA

DANUBIA was partly developed on the basis of pre-existing models – either public domain, such as MODFLOW (McDonald and Harbaugh, 1988), or proprietary developments such as PROMET-V (Schneider, 1999; Schneider and Mauser, 2000), which forms the basis of the DANUBIA Landsurface component (figure 4). The socio-economic models ("Actors" component in figure 4), however, were more often developed specifically for the use in DANUBIA, as modelling concepts for the desired purpose and scale did not exist (e.g. the WaterSupply model, see below).

As "Actors" or actor based modelling are terms which are commonly not used in hydrogeology they will be briefly explained here. Here, an "Actor" stands for any entity (or object) capable of decision making. In DANUBIA, Actors are households, farmers,

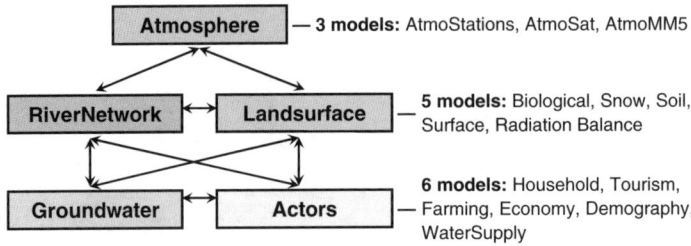

Figure 4. Connections and dependencies between the main model components and the objects (sub-models) contained in these components. Rivernetwork and Groundwater contain one model each. The "Actors" component comprises the socio-economic models (see text).

industry (entrepreneurs) water supply companies and communities. An Actor model is subsequently a model that simulates the behaviour of such Actors, e.g. the "Household Model" simulates the domestic water consumption. Actor based modelling is a term similar (but not identical) to "agent based modelling". In the actor based (socio-economic models) of DANUBIA each principal actors group (e.g. households) is split into several sub-types. Theses sub-types can have individual preferences and they react differently to changes of the outside conditions. Some further details on actors modelling in DANUBIA are provided by Janisch et al., 2006).

In the first project phase (2001–2003), research was focussed on the development of a prototype of DANUBIA (Ludwig et al., 2003). DANUBIA is a fully coupled system, comprising 16 individual models which run on different computers and exchange data at runtime using a common spatial modelling and parameter exchange concept. All model components were developed in JAVA or at least wrapped in a standardized JAVA model architecture. In DANUBIA, the individual models and their cooperation are controlled by a highly sophisticated, strictly object-oriented framework architecture developed by the computer science project group as described in Barth et al. (2004), using the graphical notation tool UML (Unified Modelling Language, Booch et al., 1999) and implemented in JAVA. Models exchange data with each other via customized interfaces that facilitate network-based parallel calculations. A prerequisite for this was the use of standardized communication procedures by all groups, such as the Java-based Remote Method Invocation (RMI). On this basis, DANUBIA was developed as a synchronized system that consists of distributed networked objects, which can communicate through RMI on the net.

The DANUBIA system has been running successfully since 2003 and is currently in a validation phase. First scenarios have been simulated, as will be shown later. The main disadvantage of the system is its computation speed (1 day computation time for 1 year simulated time on a LINUX cluster). However, within two years since the first successful run of the whole system, the performance has increased by a factor of 4.

Each discipline contributes its part of the complex model compound as an object. In this respect, an object is an encapsulated unit which completes a distinct function in the DSS and carries out the data exchange and the synchronization through defined interfaces. To minimize data traffic and to optimize the representation of complex interrelated processes, the 16 individual objects (=models in a broader sense) have been grouped to form five main model components as shown in figure 4. For example, six socio-economic models form the "Actors" component (figure 4). Common to the models in the Actor component

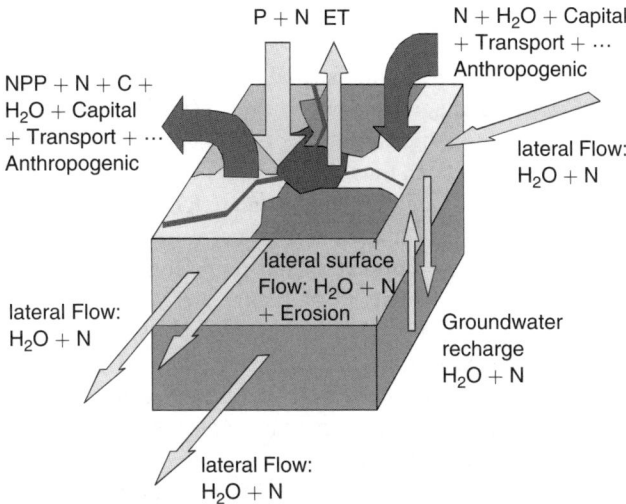

Figure 5. Schematic raster based modelling in DANUBIA on the proxel basis (Ludwig *et al.*, 2003).

are the relatively large model time steps required to model e.g. population growth, and the need to model human decisions. Furthermore, compared to the intensive data exchange within this component, data exchange between the actors component and the natural science components can be reduced to a surprisingly small number of parameters. In contrast, the Landsurface component is characterized by far faster processes and hence a shorter temporal discretisation (1 h). The data traffic and feedback within the Landsurface component is enormous. The other principal components however see little of this internal Landsurface exchange; only a few important output parameters such as groundwater recharge or nitrogen leaching are shared with models outside the component. Details on the Landsurface component are described by Ludwig *et al.* (2003).

The Institute of Hydraulic Engineering of the Universitaet Stuttgart contributes two models to DANUBIA: (1) A groundwater flow and transport model, and (2) a water supply model. They will be dealt with in more detail later on.

2.3 *Spatial and Temporal Modelling Concepts in DANUBIA*

A common problem that affects all research disciplines involved in the fully coupled system DANUBIA arises from the fact that the processes to be modelled have their main focus on different scales both in space and time. Without attempting to discuss the extensive details of such problems (few examples will be mentioned later on), it can be summarized that the different process scales can lead to undesirable feedback, incorrectness of the mass balance of the system, and, finally, to instability and low model performance. In order to avoid and overcome such drawbacks, GLOWA-Danube has agreed on a uniform spatial modelling environment and uses the concept of 1 * 1 km Process Pixels (=Proxels). Proxels are the basic building blocks of DANUBIA and consist of a pixel (picture element) in the form of a cube, in which processes occur (Tenhunen *et al.*, 1999). The proxel concept is schematically represented in figure 5. A proxel connects to its environment (neighbouring proxels) through fluxes. It can have different dimensions and layers depending on the respective processes and model concept. The standard proxel provided by the common DANUBIA architecture supplies to all disciplinary users the basic functionality for geographic referencing (e.g. ID, x, y, z) and spatial managing of the necessary parameters within the object and for data imports and exports via defined and standardized interfaces. Each disciplinary model uses a specialisation of this standard proxel which inherits all properties of the basic proxel. It can thus, for example, be a surface proxel that describes the water flow on the surface through vegetation and to the ground water proxel.

The proxel concept, which is, after all, a highly specialized extension of the traditional raster concept, is not always the optimal spatial representation for all disciplines and processes. The socio-economic models in particular are facing great difficulties when attempting to calculate quantities like GNP on a 1 * 1 km basis. However, the concept simplifies the description of the interactions in the considered interdisciplinary processes and, by being able to represent more than one dimension and sub-scale information at the same time, minimizes the disadvantages of a traditional simple raster approach.

Discretisation of time is an equally complex issue in the fully-coupled integrated model DANUBIA. Whereas a common spatial discretisation could be agreed on, model time steps must differ from model to model (15 minutes to one year). One main reason for this is that simulating very slow processes such as economic development with short times steps would result in an undesirable redundancy and low overall performance. On the other hand, processes that depend strictly on seasonally and diurnally varying parameters can not be reasonably treated using large time steps. Technically the problem of different time scales is solved using a "market place" concept. Each model puts exchange variables that are needed by another model as an input in a "public space". Making an exchange variable "public" is called "commiting". It is important to make sure that data is only committed upon the time it becomes valid and only stays "public" as long as it is valid. This is technically relatively simple but conceptually difficult if exporting and importing models simulate processes on different time scales. Depending on the individual processes aggregation and dis-aggregation of values is necessary whereby aggregation is usually simple (e.g. a monthly average of the respective diurnal values) and dis-aggregation is more difficult. How this is treated depends strongly on the sensitivity of models towards the exchange variables in question but also on feedback loops between two or even more models. A big issue is also delay caused by exchange variables that are used sequentially in different models (algorithms). Therefore, the time-related aspects are dealt with firstly by using a powerful time management tool developed by the computer science group (for details see Barth *et al.*, 2004), and secondly by a thorough joint analysis of the dynamics of coupled processes. Ludwig *et al.* (2003) exemplify this and other time related aspects in more detail. However unsolved issues remain and will be the subject of future research.

2.4 Why a Decision Support System for the Upper Danube?

The Upper Danube is a catchment with a water surplus. Hence the relevance for Global Change Research in this area is characterized less by a lack of water than by a lack of substantiated definitions of the various existing conflicts and, particularly in the Upper Danube, possible future functions in a regional management of the water resources. The natural environment in the Upper Danube is very sensitive to climate change. It is to be expected that climate change will lead to strong water- and land-use changes. However, these changes are also affected by other factors that are not related to climate change.

Among these are the creation of cultivated plants with a higher resistance to cold, precipitation, and parasites and their changed yield structure, changes in the vegetation growth and the water use efficiency due to increased CO_2 concentrations, especially at higher altitudes, and changes in agricultural production goals (quality vs. quantity) and the overall structure of agricultural industry in Germany. First impressions of this became apparent in the unusually hot and dry summer of 2003. The consequences of an average temperature of up to 6° higher than normal and 30% less precipitation were manifold. Government reports of Austria, Switzerland, and the German federal States of Bavaria and Baden-Württemberg describe the related problems in great detail. Just to mention a few, water shortages, interruption of navigation on inland water ways (Danube), problems in energy production (hydropower plants and cooling water for nuclear power plants) and severe water stress for plants and aquatic ecosystems were reported (e.g., LfU, 2004). Apart from the exceptional year 2003, tendencies of warming, decreasing glaciers, less snow cover in winter (tourism, skiing), and a shift in precipitation patterns can be observed (KLIWA, 2004; Stock, 2005).

3 GROUNDWATER MANAGEMENT IN *DANUBIA*

In the Upper Danube Basin, groundwater is the dominant source of drinking and process water (95% in the domestic, 80% in the industrial sector – without cooling). Therefore, groundwater related processes ranging from recharge, surface water in- and exfiltration, nitrogen leaching and transport, and extraction from wells for domestic, agricultural, and industrial purposes play an important role in both the physical and the socioeconomic parts of the hydrogeological cycle. It is evident that none of the corresponding processes should be treated independently, nor should they be represented in an over-simplified manner.

The management of drinking water resources in DANUBIA lies mainly in the responsibility of the research group "Groundwater Management and Water Supply" from the Institute of Hydraulic Engineering at the Universitaet Stuttgart.

Groundwater management according to worldwide or European standards such as the ones stated in the European Water Framework Directive has two main objectives: to provide water in sufficient quantity and quality to different consumers and at the same time to maintain and guarantee good qualitative and quantitative status of groundwater resources. Whereas a good quality can be described relatively simply by evaluating the chemical composition of groundwater, a good quantitative status is far more difficult to define because of the varying nature of groundwater resources of different types in different climates.

A good quantitative status of Groundwater includes firstly groundwater as a resource that should not be destroyed, depleted, contaminated and overused in order to guarantee its persistence in future times. Secondly groundwater plays an essential role for many aquatic ecosystems, in particular wetlands and meadows, and it is also the major source of river discharge in dry periods. From what has been said it is obvious that the groundwater system and its accurate representation play an outstanding role in integrated modelling systems. Within GLOWA-Danube, the research group "Groundwater" has been developing a model for the three-dimensional groundwater flow. In the integrated modelling framework of GLOWA, the groundwater model receives input from hydrological models (groundwater recharge, nitrogen leaching, river levels etc.) and delivers output to socio-economic and natural science models (groundwater level, nitrogen concentration, extraction rates etc.).

The groundwater model itself, which is described in more detail later on, is of course not "capable" of fulfilling the role of a management tool; it provides the basic parameters such as groundwater levels and fluxes but is not useful in describing technical, infrastructural, social and political aspects of groundwater management. Management after all is done by people, not by models. However, in the case of integrated models designed to simulate future scenarios, some management decisions need to be partly done within the integrated model itself. Therefore, the groundwater model provides input for a second model developed by the research group, the model "WaterSupply". Both models are integrated in the DANUBIA modelling framework (figure 4 and figure 6).

The DANUBIA object GroundwaterFlow was implemented in JAVA and wrapped around the original MODFLOW Fortran code (McDonald and Harbaugh, 1988). Interfaces exist mainly to the Actors component (withdrawal, quality), the RiverNetwork component (exchange with surface water bodies, river stages) and the Soil object (groundwater recharge, groundwater level). The object-oriented DANUBIA model WaterSupply, a member of the Actor package, is a proprietary development and was implemented entirely in JAVA. WaterSupply is in essence an interface and interpreter between the natural science models determining water supply on the one side and the socioeconomic, behaviour-driven Actors models governing demand on the other. Main interfaces in DANUBIA exist to GroundwaterFlow, RiverNetwork and the Actors objects Household, Economy, Farming and Tourism. Through a comparison of supply and demand based upon the actual organization of water extraction and distribution within the upper Danube catchment, WaterSupply aims to identify areas which may suffer water stress (Barthel et al., 2005 a).

3.1 The Danubia Groundwater Model

As mentioned earlier, the DANUBIA Groundwater component will be used to assess groundwater quantity and quality aspects of Global Change. However, in this paper only the quantitative aspects of the groundwater system and its representation in DANUBIA are covered. A transport module which is predominantly a mixed physical-conceptual approach is currently under development and will be presented upon validation.

The main aim of the DANUBIA Groundwater component is to assess and predict quantity and quality of the groundwater resources under conditions of Global Change together with the other natural science models. Commonly conceptual hydrological approaches are

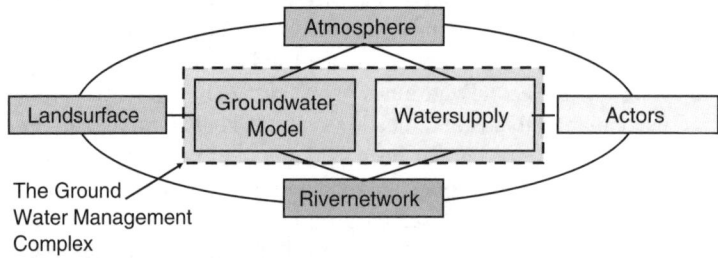

Figure 6. The groundwater management models and their relations to the main DANUBIA components. WaterSupply belongs to the Actors component but is also very closely linked to Groundwater.

used to describe the water balance of groundwater systems in large areas. However, since the distribution and change of hydraulic heads with time is an essential parameter in a coupled system like DANUBIA, a model that is capable of considering the horizontal components of groundwater flow and exchange between different aquifers is required. For example, extraction from wells should lead to a measurable local and regional drawdown in order to be able to assess environmental impacts. As a second example, nitrogen applied by farmers, and later on, leaching through the unsaturated zone, should be traceable from or to a certain drinking water well or a certain river reach. These requirements make the use of a three dimensional transient groundwater flow model inevitable. In accordance with the size of the model area and the raster-based DANUBIA approach, a finite-difference model approach (MODFLOW) was chosen. The choice of MODFLOW is also justified by the constraint to use open source models and by the proven robustness and relative simplicity of the code. Access to the source code and relative simplicity are desirable because of the need to include the code into the much larger framework of the coupled, network-based DANUBIA system. A more detailed discussion of the parameterization of the model would require a detailed description of regional and local particularities and is not feasible here. Problems of more general significance are pointed out in the following section.

The hydrogeology of the Upper Danube Catchment is characterized by four major zones: the Alps, the Molasse-Basin, the Jurassic Karst (plus other Mesozoic rocks) and the Crystalline Basement Complex (figure 7). The folded, thrusted and faulted Alps make up about 30% of the region. In the Molasse Basin north of the Alps, unconsolidated to semi-consolidated clastic sedimentary formations predominate. To the northern and north-eastern boundary, the basin is surrounded by sedimentary and crystalline rock formations. The Jurassic Karst dominates the hydrological situation at the northwestern part (figure 7). The hydrogeological conditions vary extremely both in horizontal and in vertical direction. None of the dominating aquifers exists all over the domain. In the alpine part, which is especially difficult to integrate in a deterministic groundwater flow model, large continuous aquifers are completely unknown. One main challenge in setting up the conceptual model for the heterogeneous catchment is to find the appropriate number and extent of aquifers needed to describe the main flow characteristics of the basin and to create meaningful model output. In view of the requirements of the integrated system, the data availability, the computation time needed, the stability of the numerical model and of course the obligatory discretisation of 1 * 1 km cells, the final concept is the result of iterative process and includes many compromises. In the integrated system the main focus is on the coupled processes close to the land surface, such as groundwater exchange with surface waters, soil, biosphere and atmosphere, and on the exchange with the human part of the water cycle, i.e. water consumption and contamination. This has lead to a conceptual model that focuses rather on the shallow parts of the groundwater system and short to medium term processes. Deep flow systems and long-term processes are neglected due to their minor contribution to the actual water cycle.

The conceptual model consists of four layers, comprising the strata "Jurassic Karst", "Younger Tertiary", "Older Tertiary" and "Quaternary" (figure 8). Only aquifers with basin-wide occurrence are considered due to insufficient data availability, the model grid resolution, and requirements of the MODFLOW approach. These four modeled layers are not always present (active) over the entire domain. This is especially problematic for the thin network of alluvial aquifers and gravel plains of the lowlands (figure 7 and figure 8). These aquifers are very important for the exchange of groundwater with surface waters

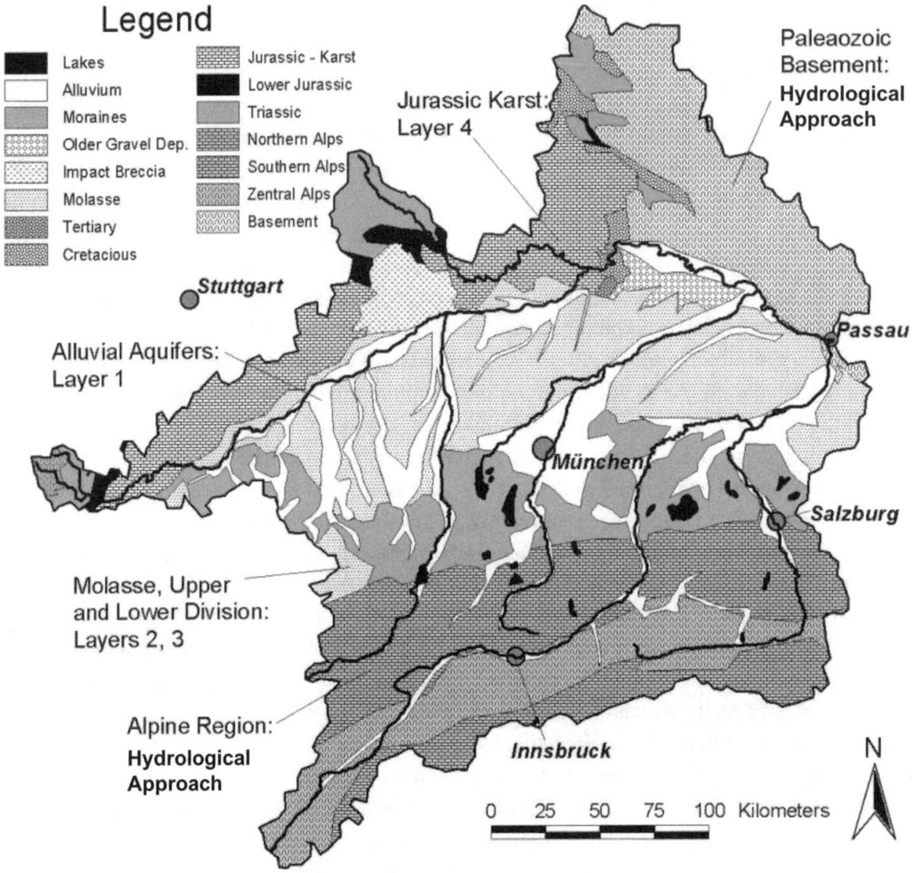

58   Groundwater flow understanding from local to regional scale

Figure 7.   Schematic geological map of the upper Danube basin.

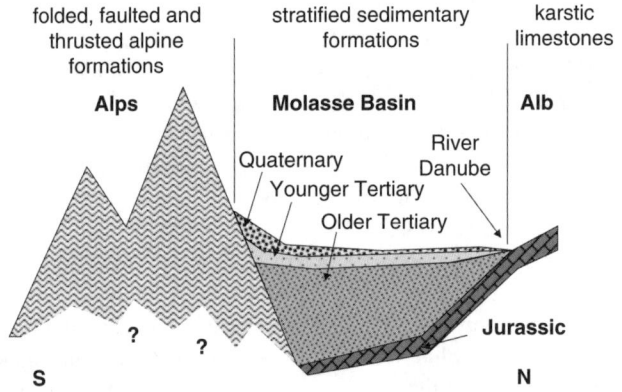

Figure 8.   Schematic geological cross section of the upper Danube basin showing the four model layers.

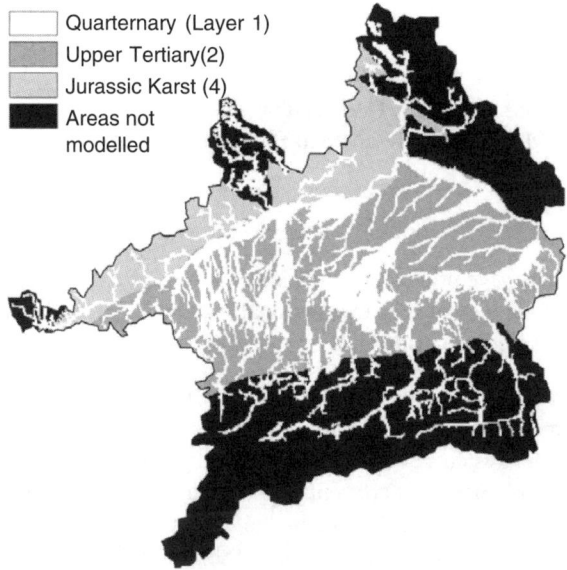

Figure 9. Horizontal distribution of the uppermost active layer of the Groundwater model. Please note the complex geometry of the Quaternary layer.

and the atmosphere and for water supply. The Quaternary layer is mainly defined by small and thin local structures of high permeability (valley aquifers, alluvial gravel plains), which are of high importance in DANUBIA as an integrated system. Taking place within the alluvial aquifers is the major part of the groundwater-surface water exchange, the link to groundwater dependent ecosystems (swamps, wetlands, meadows), evaporation and plant water uptake from groundwater and last but not least groundwater extraction for drinking water purposes. Unfortunately, the complicated geometry of the Quaternary layer makes it the most difficult to model. Importance and complexity of the layer brought it into the centre of research activities.

3.2 Integration of MODFLOW in DANUBIA – adaptation of the PROXEL concept

Parallel to the development of the Groundwater model, the integration of this model in the structure of DANUBIA was pursued. The finite-difference model MODFLOW was chosen mainly because of the cell-based approach that matches the proxel concept of DANUBIA in a nearly ideal way. One to one data exchange with other models is possible without elaborate post-processing of the model output. Although the block-centred flow approach used by MODFLOW has numerous advantages (simplicity, robustness, perfect integration in DANUBIA), it also has clear disadvantages, particularly with regard to the implementation of boundary conditions and the representation of complex geometrical features.

The following input data are calculated by the models (figure 4) named in parenthesis: river level (RiverNetwork), nitrogen in surface water (RiverNetwork), groundwater recharge (Landsurface), nitrogen in percolating water (Landsurface), groundwater withdrawal (WaterSupply). Likewise, the following output data are required by the models stated in parenthesis: groundwater level (Landsurface), nitrogen in groundwater (Landsurface, RiverNetwork, WaterSupply), and infiltration and exfiltration between groundwater and surface water (RiverNetwork). The transfer parameters were implemented in UML-diagrams, which in turn were used to create a JAVA code which can be integrated in the overall structure of DANUBIA. More details are given in Barthel *et al*. (2005 b).

3.3 Crucial aspects of groundwater flow modelling on a very large scale

Three major problems have proven to be decisive in the attempt to successfully model the groundwater flow dynamics of a coarse regional groundwater model with complex geological conditions on a coarse grid:

(a) The appropriate representation of the complex aquifer geometry on a coarse grid required some manual adjustments to aquifer size and extent. After the identification of the regional aquifer systems and the creation of a hydrogeological conceptual model, it is important to implement this concept into the groundwater model such that a stable numerical solution of the model is attainable. The main problem is to achieve a connected aquifer system which is able to receive the groundwater recharge in the mountainous areas (Alps to the South, figure 7, figure 8) and which yields a reasonable base flow at existing gauging stations in the forelands. Due to the discrepancy between the finite difference cell size and the extent of the narrow, highly permeable aquifers, additional highly permeable cells have to be "added" in order to achieve a close solution for groundwater flow using a finite difference scheme. In addition, it has to be ensured that each cell of this "virtual" aquifer has at least one neighbouring cell (in the direction of groundwater flow) with a lower base to guarantee the conductivity of the aquifer. The concept just briefly described was used to implement an algorithm that allows the detection of cells whose permeability needs to be adjusted and to add cells to the modelled aquifer layer. The algorithm was applied to the catchment of the Upper Danube (for details see Wolf *et al*., 2004). The modelling results of a finite difference groundwater model in this area using an adjusted aquifer geometry are very promising. Measured groundwater levels in the gravel aquifer can be modelled with an accuracy of less than two meters (figure 12, figure 13). Without a proper investigation of the regional aquifer system and the application of the presented algorithm for the discretisation of such a system, the modelling of regional groundwater flow on a coarse finite difference grid would not be possible at all.

(b) In the Alps and a crystalline region in the Northeast of the basin (figure 7, figure 8), only small, disconnected saturated zones exist. Groundwater flow is restricted to fracture zones or karstic systems, which under the given constraints, and because of missing data, cannot be included in a regional model. The alpine section of the model area is a subject of particular concern. On the one hand, the alpine regions, covering approximately 30% of the catchment area and contributing about 40–50% of the total precipitation, evidently play a major role in the water cycle of the region. On the other hand, it is not possible to treat the extremely faulted, folded, and thrusted stratigraphic units of the Alps as ordinary quasi-horizontal layers as they are usually described in the MODFLOW concept. Different alternative modelling approaches are available (explicit description of fracture and matrix flow, double-porosity model, etc.), however their implementation poses difficulties, either because the theoretical foundations of the method are still in development, or simply because of the lack of data needed to parameterise the model. As one cannot extrapolate point data to obtain area information (as it is done for porous media), no direct (measured or estimated from measurements) quantitative assessment of the effective parameters characterising groundwater fluxes through rock masses can be performed. In order to overcome the problems described above, a combined deterministic-conceptual approach was developed and implemented. In this approach, the finite-difference Darcy law based model was extended to its maximum validity domain, namely to the alluvial aquifers draining the water from the mountains into the foreland. For the rest of the area, separated into hydrological sub-catchments built on the base of the digital elevation model, only qualitative conceptual hydrological models were developed. At the moment, a process oriented approach including a joint calibration of the groundwater flow model for the valley regions and the hydrological model for the alpine parts has only been implemented for a number of sub-catchments (Rojanschi *et al.*, 2004). In the "large" model, a simplified approach is still used. Here the groundwater recharge of each sub-catchment is routed without temporal delay to a pre-defined model cell in the valley aquifer. This temporary solution but will soon be replaced by the aforementioned approach.

(c) In a coupled regional model, the groundwater recharge, commonly defined as the amount of water percolating through the plant-influenced soil zone, has to be determined considering the processes in the deep unsaturated zone, where horizontal unsaturated/saturated flow can predominate. In sub-domains that are characterized by very deep regional groundwater tables, or deep confined aquifers, perched aquifers, which cannot be modeled in a regional model, predominate in the uppermost part of the subsurface (0–200 m). On the regional scale, local perched aquifers have to be treated as part of the unsaturated zone. Horizontal flow leads to discharge of percolating water in springs and small tributaries. It has proven to be extremely difficult to determine the actual recharge to the deep groundwater system, in particular because data to describe this deep, partly saturated zone does not exist. The approach to tackle this problem is to regionalize the factors that determine the amount of horizontal discharge and the deep percolation rates of the deep, unsaturated zone, and use them to parameterize the corresponding transfer functions. However, no satisfactory solution to this problem is available yet, and the authors are convinced that a considerable amount of basic research on all scales is still required.

3.4 The DANUBIA Watersupply Model within the Actors component

As stated in a previous section, the WaterSupply Model is the part of the Groundwater Management complex (figure 6) of DANUBIA which represents the technical, infrastructural and human aspects of management. It draws on physical parameters calculated by the groundwater and surface water models and interacts in various ways with other parts of the DANUBIA system (figure 4), as described below. Since water supply models of that kind are new (to the knowledge of the authors) their purpose and role will be explained in a little more detail. An extensive description can be found in Barthel *et al.* (in press a). The development of the water supply model entailed a careful consideration of the following aspects:

- What are the general aims and the common concepts of the integrated system DANUBIA and how can the specific purpose and role of the WaterSupply object within the integrated system be defined?

- What are the possible self-contained aims and specific, field-related problems the model might strive to answer which may be of interest independent of the integrated system?
- What are the relevant elements of a water supply system and the relevant processes to be included and, in light of future scenarios, how much dynamic or, in other words, to which extent should which parts and elements of the model be able to react to changing boundary conditions?
- What are the boundaries and area of expertise of the model and the parameters to be exchanged between WaterSupply and other models, and what modelling technique and degree of sophistication are appropriate?
- And, very important, what data is required to be able to model the elements and processes found to be important and what is the actual availability of data?

Within DANUBIA, WaterSupply must carry out the following functions:

- Relay the spatially and temporally variable water use of the various Actor models to Groundwater or Rivernetwork at the appropriate geographic location;
- Interpret the spatially and temporally variable state of the water bodies (quantitative and qualitative),
- Model the decision for and implementation of technical measures to ensure that the total demand for drinking water can be met if at all possible or otherwise conveys the degree of necessity to limit use to the Actor models;
- Calculate the water price, using a function to be developed in a joint effort with the "Economy" group.

These aims were approached by focussing upon the organizational structure of water supply in the Danube basin, which shows a complex, intertwined hierarchy of public water suppliers. Characteristic of water supply within the upper Danube basin is a groundwater-dependent, strongly decentralized, three-tier structure comprising local, community based suppliers (well over 2,000), regional special purpose associations (~300) assuming the water supply responsibilities (maintenance, administration, financial matters, and in many cases also technical infrastructure) for a group of communities, and a few supra-regional, long-distance suppliers (~5) supplying regions with few or no resources (Emmert, 1999). Although the use of local resources is generally preferred, many communities draw upon supply from all three organizational forms for security purposes. A number of group suppliers and in particular the long-distance suppliers import or export appreciable amounts of water across the boundaries of the Danube basin, which need to be accounted for in the water balance.

Due to the common use of an object-oriented modelling approach and the focus upon a central "actor" (here the water supply companies WSC), WaterSupply is a member of the actor group (figure 4). In contrast to the other actor models, it is not the declared aim of WaterSupply to predict the decisions and actions ("behaviour") of the WSC in response to changing climatic, demographic, technological, etc. conditions. Rather, the identification of the possibility as well as need for action (meaning water availability and water quantity) is at the centre of attention.

Currently, great efforts are being made to further merge the GroundwaterFlow and the WaterSupply objects in order to fulfil the task of creating an integrated tool for Groundwater Resources and Supply Management. This is especially important for the "Deep Actor Model" deepWaterSupply, which was just successfully implemented. Deep Actor Models are comprised of a number of individual "Actors", objects which perform different actions depending

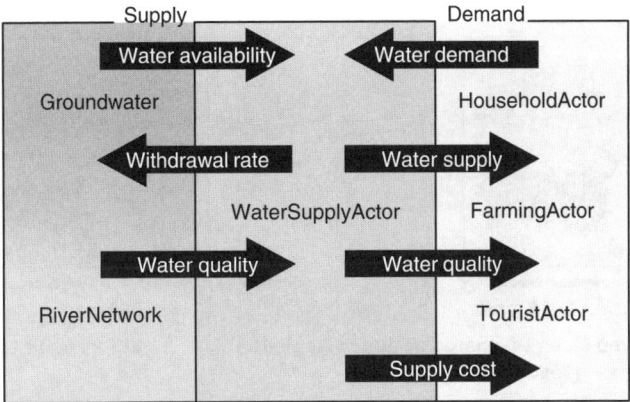

Figure 10. Water supply: the interface between supply and demand.

on their individual attributes. A common Deep Actors architecture or framework, similar to the common DANUBIA framework, is used to model decisions similarly in all Deep Actor Models. In DeepWaterSupply, the central actors are the water supply companies. The WSC objects decide on specific plans and actions based on analyses of parameters calculated by the Groundwater, Rivernetwork and Landsurface components. The output of the latter is here used to calculate resource availability using key parameters that describe the state and trends of groundwater resources. The utilization of the Deep Actors concept leads in case of the WaterSupply Model to a far more flexible and realistic treatment of the sustainability problem.

4 DANUBIA MODEL RESULTS

Figure 11 shows a summarized water balance (based on only two output parameters) from a DANUBIA validation run (1970–2000) that shows that the system as a whole works reasonably well at least if one looks at the natural science model components and basin-wide long term results. More detailed result descriptions for DANUBIA as a whole can be found in Strasser *et al*. (2005). In the following, disciplinary results related to groundwater management will be discussed in more detail.

The current working version of the groundwater model has been successfully run and tested within the DANUBIA environment after careful adaptation of the model geometry, parameter upscaling and calibration for both steady state and transient conditions.

The piezometric heads calculated by the DANUBIA Groundwater component are generally acceptable when compared to measured mean values (overall $R^2 = 0.97$ for steady state results, figure 12) and time series (figure 13). However, big differences exist in various parts of the basin and for various aquifer sections. Generally, the deviations from the natural situation are small for the unconsolidated, quaternary aquifers that fill river valleys and gravel plains (layer 1, figure 8, figure 9) but large for the Jurassic Karst, the Alps, the crystalline regions and parts of the Tertiary (figure 12). Since the Quaternary aquifer is the most important for the short to medium term (days to several years) exchange of the groundwater with surface water bodies and the atmosphere, this is in many cases acceptable.

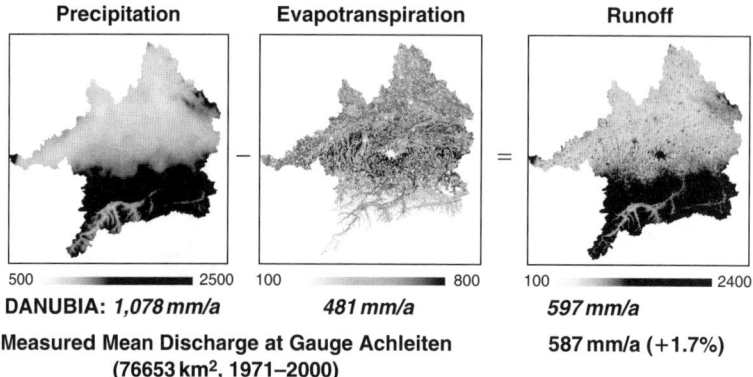

Figure 11. Water Balance of the Upper Danube, Period 1971–2000.

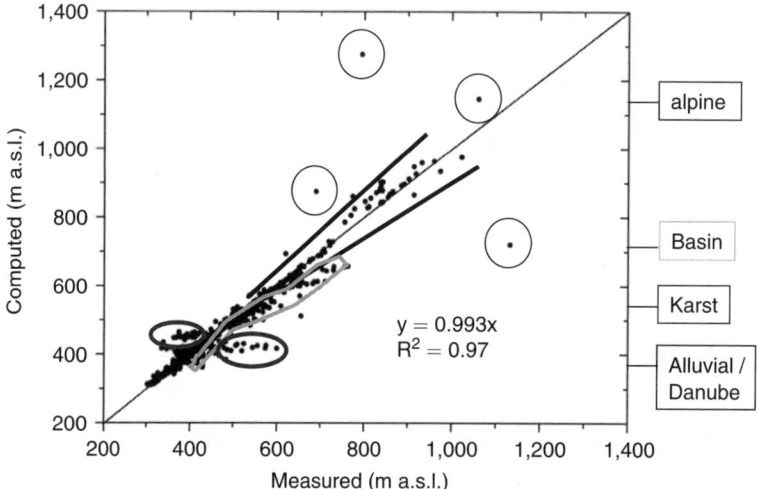

Figure 12. Comparison of measured and calculated values of groundwater heads for a steady state simulation.

A comparison of the model results calculated for a reference period (1995–1999) and a wet (95, 96 + three times 2002) and a dry scenario (95, 96 + three times 2003) shows that the model reacts in a reasonable way to the main input parameters, namely the groundwater recharge calculated by the DANUBIA soil model. This is the case for single model cells (figure 13) as well as for the whole catchment (figure 14). In both cases a significant decrease of the groundwater level results from the lower groundwater recharge (30% less) originating from the much dryer climatic conditions in the exceptionally hot and dry year 2003. Figure 13 reveals a limitation of the large scale model, namely the smoothing effect of the relatively large grid size of 1 * 1 km. Depending on the nature of problems the model will be used to solve, this has to be accounted for.

Integrative modelling of global change effects 65

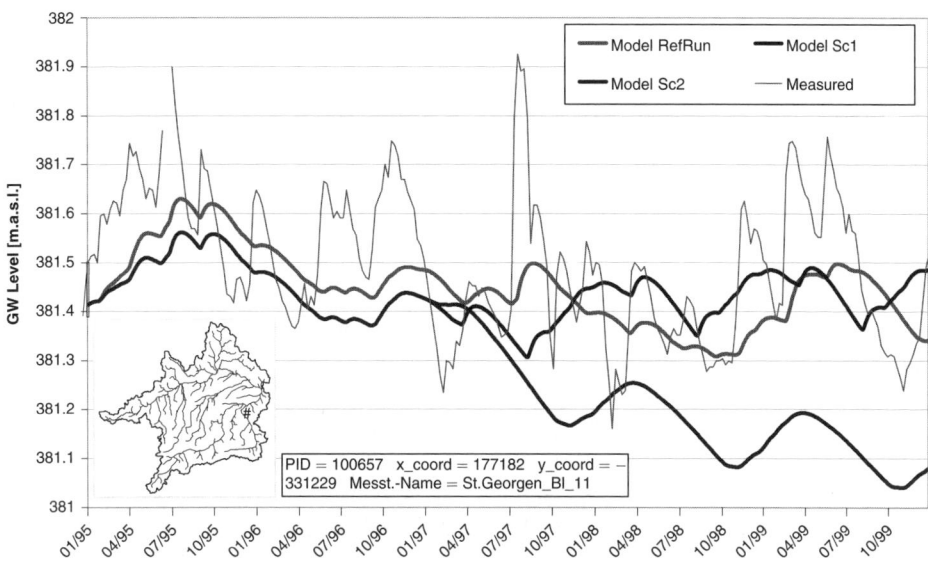

Figure 13. Comparison of measured and calculated groundwater levels for an observation well located near the river Salzach close to the Bavarian / Austrian border. RefRun: Model validation period 1995–1999, Sc1: "wet" scenario, 1995–1996, 2002, 2002, 2002; Sc2: "dry" scenario, 1995, 1996, 2003, 2003, 2003.

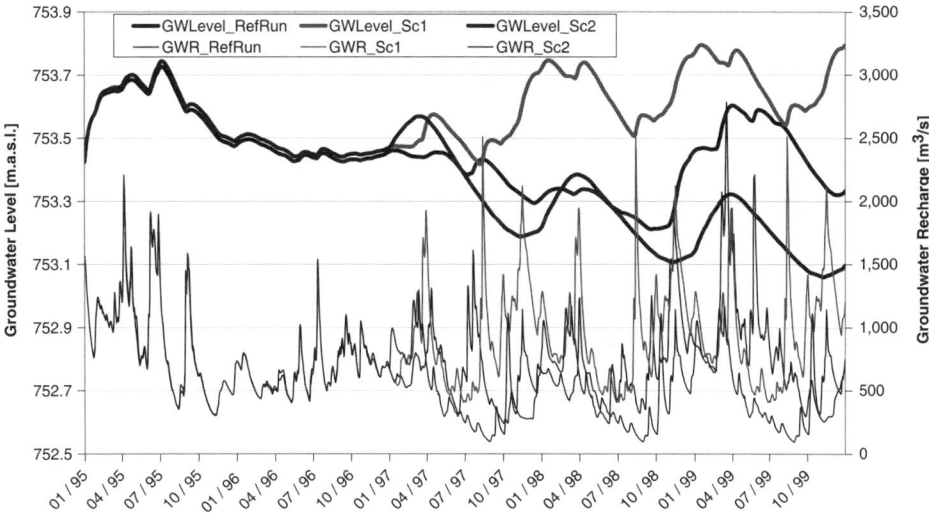

Figure 14. Comparison of the mean groundwater level and the mean groundwater recharge for the whole catchment for the reference period (RefRun), the wet (Sc1) and the dry scenario (Sc2).

The scenario results for the DANUBIA WaterSupply model are, as for all actors models, less significant due to the relatively short simulation period of five years. However, it can been seen in figure 15 that the domestic drinking water demand increases noticeably during the hot and dry summer in 2003. This, in turn, resulted in a slightly higher total groundwater

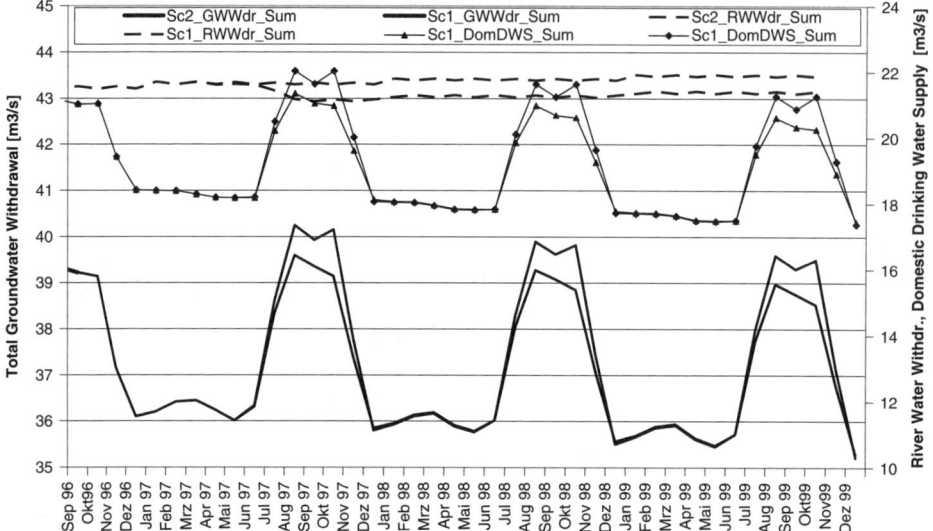

Figure 15. Comparison of domestic drinking water supply, groundwater and river water withdrawal for the whole catchment for the wet (Sc1) and the dry scenario (Sc2).

withdrawal, which, however, was negligible looking at the overall water balance of the groundwater flow component. In addition, the decrease in groundwater recharge plus the increase in water demand did not yet invoke a limitation in drinking water supply. Since such extreme conditions were not known in the Upper Danube catchment in the past, it is difficult to decide how to set the thresholds. This will in future be discussed with different stakeholders.

In order to understand the WaterSupply model results, it is important to remember that the main role of WaterSupply is to act as a link between the demand and the resources side of the systems. Being a linking part in an integrated system, the "results" are highly dependent on the results of the connected models. A "good result" of WaterSupply is achieved if all the demands can be satisfied following the predetermined patterns of water distribution patterns in the real world. From the decision makers point of view, interesting results are only to be expected in cases where the present day situation, which is characterized by an almost 100% satisfaction of demands is disturbed, e.g. by extreme climatic conditions. Only then will the "business as usual" mode of behaviour be left. Such deviations can then be interpreted.

In figure 16 the difference between the DomesticDrinkingWaterDemand calculated by the Household Model (figure 4) and the DomesticDrinkingWaterSupply calculated by the WaterSupply Model is shown for a winter and a summer situation in 1999. The results originate from a simulation used for model validation for the years 1995 to 2000. All models were previously tested and adjusted for the years 1995 and 1996. Since the input values were slightly different in 1999 from the 95/96 values that the model was adjusted to, a deficit for a small number communities was calculated (15 in winter, 45 communities in the summer). However, no water scarcity is known for 1999. On the other hand, the deficits are very small and the percentage of undersupplied communities is less than 1% or 2% for the winter and summer respectively. Nevertheless a deficit in 1999 has to be considered an error.

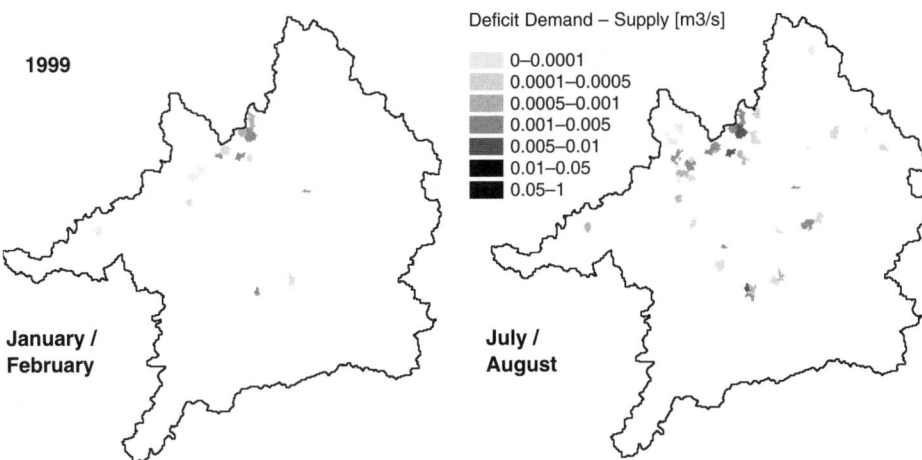

Figure 16. Deficits for 2,133 communities in the Upper Danube Catchment calculated in a 1995–2000 validation run of the integrated system DANUBIA based on the DomesticDrinkingWater Demand and the DomesticDrinkingWaterSupply for January (left) and July (right) in 1999. In the coupled system a supply calculated in time step 2 is based on a demand calculated in time step 1. Time step length for this simulation was one month.

To find the cause of this error is a difficult task in the integrated system because it might not even be an error in the WaterSupply model (its data base or algorithms) but also an error in the partner models, e.g. the Atmosphere model that calculates the precipitation which is used to calculate the groundwater recharge in the soil model and so forth.

5 CONCLUSIONS

After more than four years of development of the DSS DANUBIA in a large interdisciplinary research consortium, a consistent, meaningful, technically working solution can be presented. Using a network-based approach and the discipline-independent diagrammatic modelling language UML, the technical and scientific basis of the decision support system DANUBIA was designed and implemented during the first phase of GLOWA-Danube (2001–2004). It considers all hydrological and many socio-economic processes related to the water cycle. The results of the simulations are now being presented to the relevant stakeholders. Among the stakeholders are members of the water management authorities of the different political-administrative entities, the agricultural management authorities, the power industry, and the tourist boards. Appropriate scenarios for further simulation runs will be developed in cooperation with them. Developing the basin-scale groundwater model proved to be a challenge, but, after careful consideration of the model geometry and parameterisation, a solvable problem. The results, however, should always be regarded as results of a regional model, lacking the spatial and temporal details of local simulations.

With regard to the groundwater related parts of the integrated system, important steps towards a better understanding of groundwater modelling on the very large scale were made. The upper basin is very complex with respect to geology and hydrogeology. Any experienced groundwater modeller will agree that under such conditions, meaningful groundwater flow modelling is very difficult. However, the feasibility depends on the nature of the desired results. Integrated Water Resources Management, especially on the catchment scale, depends usually on regional, rather than on local results. Predictions are needed for long term processes rather than detailed descriptions of short term variations. In addition we found, that volumes and fluxes (such as baseflow) are much more important in integrated systems than piezometric heads, because the former are the quantities that are primarily of interest in water resources management.

The question as to how much data is necessary to set up reliable models is a very interesting one. On first sight it seems obvious that one can never have enough data (parameters like hydraulic conductivity, porosity storage coefficients). This proves to be only partly true.

More important than hydraulic parameters is data to construct the appropriate model geometry. The model geometry in a geologically and topographically complex region is a difficult matter, especially if one deals with a relatively coarse discretisation. The key issue is to guarantee flow and to avoid "bottle necks", both extremely difficult if aquifers are thin and shallow. Thin aquifers situated diagonal to the principal flow direction in areas with steep relief and steep groundwater gradients are prone to be flow obstacles. Therefore the adjustment of aquifer bottom and top, groundwater level, river level and the land surface can often only be achieved by artificially adapting the aquifer geometry. However, setting up the model geometry from scratch, i.e. from borehole data, is nearly impossible. If (digital) contour maps of significant boundaries do not exist, the construction of the model geometry for such large and heterogeneous areas becomes a very difficult task. It is possible that a different modelling concept, e.g. finite elements, would be more appropriate to solve the geometrical problems mentioned here. Different constraints such as the requirement to use public domain, open source software finally led to the decision to use the finite difference approach (MODFLOW) accepting the inherent discretisation difficulties. In any case the model domain remains large and complex – a challenge for all groundwater flow modelling concepts.

The hydraulic model parameters are surprisingly enough of minor importance. This has two reasons: It is only possible to upscale field measurements to a grid size of 1 * 1 km for limited cases. The degree of heterogeneity in both vertical and horizontal direction for such a coarse discretisation is so large, that, even if enough fine resolution data would be available, no meaningful effective parameters can be calculated (Rojanschi, 2001). Using rough estimates, general geological knowledge and mean values from literature has proven to be sufficient.

For model calibration and adjustment piezometric head data is very important. Here it proves to be problematic that head data mainly exists where groundwater is actually used, i.e. for certain aquifers in densely populated areas. For vast sections of the catchment and for many of the deeper aquifers, not a single observation well exists. In addition, it is not always clear which aquifer is filtered in an observation well. Still, the problem remains that on a coarse grid, a measured groundwater level cannot be compared with a modelled groundwater level of a cell without further considering topography and heterogeneity of the same cell.

It would be desirable to know much more about effective groundwater recharge (the part of the recharge that actually reaches the aquifers being modelled), interflow, base flow and other immeasurable quantities. They are related to the most important boundary conditions (fluxes in and out) and determine to a large extent the performance of the groundwater model. Unlike the traditional groundwater modelling, in our integrated system the recharge is calculated by other groups, and in turn the infiltration to surface waters (baseflow) is used by other partner models. To close the balance the groundwater model has to be adjusted to both inflow and outflow. Neither can be neglected nor controlled by the groundwater modeller. This is a relatively unusual situation for the groundwater modeller who is usually not responsible for completely closing the whole water balance between the clouds and the outlet of a catchment. In integrated models all parties share this responsibility.

Finally, it is a fact that today the data needed to set up such large models is not readily available from one data source. It has to be gathered from many different locations, administrative bodies and literature. It must be digitized and homogenised. In our case obtaining and preparing the data consumed much more time than running and calibrating the model.

In summary, data is very important as always in groundwater modelling. But, in contrast to traditional small scale groundwater models, measured data becomes less important for the coarser discretisation as the up-scaled parameters become less meaningful. The physical meaning of a K-value of a 1 * 1 km, 100 m thick model cell in a heterogeneous geological environment is very limited. It is however important to define top and bottom of this cell in a way that respects the natural conditions but at the same time guarantees numerical stability.

As described in the previous sections, modelling groundwater flow on the large scale in integrated systems is challenging. Data availability plays an important, but not the foremost role. Despite the difficulties mentioned here it has to be said that groundwater models of this size and heterogeneity can be applied successfully if two main aspects are considered carefully: A lot of effort should be put in setting up the appropriate conceptual model, namely in the adequate definition of the model layers geometry and boundary conditions. Secondly one should always keep in mind that such models, especially as a part of IWRM system should only be used to address long term, regional problems. An application to local, short-time period questions is not allowed. Here the large scale model can provide boundary conditions for smaller, (nested) high-resolution local models

The development and validation of an object-oriented water supply model for the upper Danube area with the functionality described above represents a milestone in the project GLOWA Danube. The model in its present form will remain a part of the DSS DANUBIA since it is now capable of fulfilling its role in the desired circumference. However, after the end of the first phase of GLOWA-Danube the research cooperation defined new goals that also affect the further development of the WaterSupply. In the second project phase (2004–2007), the focus will shift towards the active integration of the stakeholders from the field of water resources management. Decision-making "rules" will be debated with the relevant stakeholders and adapted where necessary. Based on these rules, the object-oriented WaterSupply component will be transformed to a deepActor model with limited decision-making functionality as described earlier on, with WSC actors able to respond to their environment and behave in a goal-oriented manner to bring about change in the water supply system in response to changing conditions with regard to the climate, water availability and quality, political and social boundary conditions, and changing demand. The second project phase will furthermore be dedicated to the refinement of the various GLOWA-Danube models and to the formulation, testing and comparison of complex scenarios of future development with the aim of identifying sustainable forms of water management and consumer behaviour. Ultimately, DANUBIA will be able to serve as a tool for monitoring, analysing and modelling the impacts of Global Change on nature and society in the Upper Danube basin for various future scenarios, taking into account a multitude of environmental, social and economic aspects formulated by the water-related stakeholders.

Groundwater management is a task that has to be worked on in an integrative and interdisciplinary way. A coupled system as DANUBIA forms a perfect platform to do this since it covers all the natural science and socio-economic aspects involved. However, one should not mistake a system like DANUBIA as a traditional management tool used to solve a site-specific, well-defined problem. DANUBIA, and its Groundwater and WaterSupply models in particular, can only address regional problems with a long-term perspective. It is meant to answer questions related to Global Change and its regional consequences. The temporal perspective is 30 to 100 years, the spatial domain of validity is the whole catchment or large parts of it. When evaluating and assessing DANUBIA and its results, that has to be taken into account.

ACKNOWLEDGEMENT

The authors wish to thank the German Ministry for Education and Research, the Bavarian State Ministry for Science, Research and Fine Arts, the Baden-Württemberg Ministry of Science, Research and the Arts and the Universities of Munich and Stuttgart for funding GLOWA-Danube. Furthermore we like to thank all partners in GLOWA-Danube for their support. Especially we would like to thank Darla Nickel for critical review of the manuscript.

REFERENCES

Barth M, Hennicker R, Kraus A, Ludwig M (2004) DANUBIA: An Integrative Simulation System for Global Change Research in the Upper Danube Basin. Cybernetics and Systems, Vol. 35 (7–8), pp. 639–666.

Barthel R, Nickel D, Meleg A, Trifkovic A, Braun J (2005 a) Linking the physical and the socio-economic compartments of an integrated water and land use management model on a river basin scale using an object-oriented water supply model. – Physics and Chemistry of the Earth.

Barthel R, Rojanschi V, Wolf J, Braun J (2005 b) Large-Scale Water Resources Management within the Framework of GLOWA-Danube – Part A: The Groundwater Model. – Physics and Chemistry of the Earth.

BMBF (German Ministry of Research and Education, 2002) German Programme on Global Change in the Hydrological Cycle Status Report 2002 (Phase I, 2000–2003) (http://www.glowa-danube.de/PDF/reports/statusreport_phase1.pdf).

BMBF (2005) GLOWA Global Change in the Hydrological Cycle – Status Report 2005, 158 p, Bonn, Germany http://www.glowa-danube.de/PDF/reports/statusreport_phase2.pdf

Booch GJ, Rumbaugh I, Jacobson I (1999) The UML User Handbook, Addison-Wesley, San Francisco, U.S.A.

Emmert M (1999) Die Wasserversorgung im deutschen Einzugsgebiet der Donau. Wasserwirtschaft 89, 7–8, 396–403.

Janisch S, Barthel R, Schulz C, Trifkovic A, Schwarz N, Nickel D. (2006) A Framework for the Simulation of Human Response to Global Change – Geophysical Research Abstracts, Vol. 8, 06195, 2006.

KLIWA (2004) Klimaveränderungen und Konsequenzen für die Wasserwirtschaft – 2. KLIWA-Symposium Würzburg. KLIWA-Berichte Heft 4. ISBN 3-937911-16-2.

LfU (Landesanstalt für Umwelt, 2004) Das Niedrigwasserjahr 2003 – Oberirdische Gewässer, Gewässerökologie 85, ISSN 1436-7882 (Vol. 85, 2004). http://www.lubw.baden-wuerttemberg.de

Ludwig R, Mauser W, Niemeyer S, Colgan A, Stolz R, Escher-Vetter H, Kuhn M, Reichstein M, Tenhunen J, Kraus A, Ludwig M, Barth M, Hennicker R (2003) *Web-based modelling of energy, water and matter fluxes to support decision making in mesoscale catchments-the integrative perspective of GLOWA-Danube.* – Physics and Chemistry of the Earth 28, pp 621–634.

Mauser W, Barthel R (2004) Integrative Hydrologic Modelling Techniques for Sustainable Water Management Regarding Global Environmental Change in the Upper Danube River Basin. – In: Research Basins and Hydrological Planning (Ed., Xi, R.-Z., Gu, W.-Z., Seiler, K.-P.), pp. 53–61.

McDonald MG, Harbaugh AW (1988) A modular three-dimensional finite-difference ground-water flow model: *U.S. Geological Survey Techniques of Water-Resources Investigations, book 6*, chap. A1, Washington, U.S.A.

Rojanschi V (2001) Effects of Upscaling for a Finite-Difference Flow Model. – Master's Thesis, Institut für Wasserbau, Universität Stuttgart.

Rojanschi V, Wolf J, Barthel R, Braun J (2004) Considerations about the Integration of Deep Unsaturated Zones and Rock Formations into Hydrological Distributed Models of Large Mountainous Catchment Areas. Geophysical Research Abstracts, EGU – 1st General Assembly (25–30 April 2004, Nice, France), European Geosciences Union, 4/2004. – Vol. 6.

Schneider K (1999) Spatial Modelling of Evapotranspiration and Plant Growth in a Heterogeneous Landscape with a coupled Hydrology-Plant Growth Model utilizing Remote Sensing. In: Proc. Conference Spatial Statistics for Production Ecology, Vol. 4, Wageningen.

Schneider K, Mauser W (2000) Using Remote Sensing Data to Model Water, Carbon and Nitrogen Fluxes with PROMET-V. In: Remote Sensing for Agriculture, Ecosystems and Hydrology, SPIE Vol. 4171, S.12–23.

Stock M (Ed.) (2005) Klara, Klimawandel – Auswirkungen, Risiken, Anpassung. – Pik Report, No 99.

Strasser U, Mauser W, Ludwig R, Schneider K, Lenz V, Barthel R, Sax M (2005) GLOWA-Danube: Integrative Global Change Scenario Simulations for the Upper Danube Catchment – First Results. In: Papers of the International Conference on Headwater Control VI: Hydrology, Ecology and Water Resources in Headwaters. IAHS, Bergen, Norway.

Tenhunen JD, Kabat P (Ed.) (1999) *Integrating Hydrology, Ecosystem Dynamics, and Biogeochemistry in Complex Landscapes*, John Wiley & Sons, Chichester, 1999.

Wolf J, Rojanschi V, Barthel R, Braun J (2004) Modellierung der Grundwasserströmung auf der Mesoskala in geologisch und geomorphologisch komplexen Einzugsgebieten. – Tagungsband zum 7. Workshop zur großskaligen Modellierung in der Hydrologie, November 27–28, 2003, LMU München.

CHAPTER 2

Water management in transboundary hard rock regions – A case study from the German-Czech border region

S. Bender, T. Mieseler, T. Rubbert and S. Wohnlich
Chair of Applied Geology, Ruhr-University Bochum, D-44780 Bochum, Germany

ABSTRACT: Border regions, which are geologically characterised by hard rocks, combine the disadvantages of limited groundwater reservoirs and an unfavourable geographical location. In order to optimise water management in borderland regions, the pilot area Šumava (PA Šumava) was chosen as an example for cross-border groundwater flow in granite and gneis regions. Furthermore, within the TRANSCAT-project this PA was selected due to a high degree of groundwater protection with little anthropogenic impact only, giving it a particular position among the five PAs. By a close bilateral co-operation between Germany and the Czech Republic it is planned to identify key indicators and standardised methods for data acquisition providing comparable data sets to be implemented in the TRANSCAT-Decision Support System for optimum water management in borderland regions.

Keywords: Water management, hard rock, DSS, Bohemian massif, transboundary, Transcat.

1 INTRODUCTION

In most countries, there exist legal frameworks of laws, norms and regulations which define standard procedures for sampling and analytics. This legal framework differs from country to country in certain terms, but a basic level of conformity is given by norms which are applicable EU- or even worldwide, as EN- and ISO-norms. Still, there are many regional and national guidelines or limit values for environmental parameters, so quality aspects can change rapidly when crossing a border. Furthermore data sets are different depending on the part of the border region due to different campaigns of sampling and analytics existing. While for example in the German part of PA Šumava, "Region Upper Regen", there are smaller parts which have been intensively investigated in other research projects (around the city of Bodenmais for example over 30 springs have been qualitatively monitored for several months), in most parts of the nature parks "Bayerischer Wald" and "Šumava" data exists only at few sampling points. Even more important than the amount of available data are data quality and comparability of information from different data sets. It has been shown that on-site-monitoring may be suitable in some cases to obtain high-quality data, while under certain circumstances it may be not. Furthermore, the definition of main indicating parameters for each pilot area is as important for an operative Decision Support System as the consideration of scaling effects and sampling intervals (Bender *et al.*, 2004a). Water management in Southern Germany is mainly focused on groundwater regions with predominating porous aquifers, such as the hydrogeological region of the Alpine foothills moraine belt. Their groundwater yield is high due to a combination of high precipitation rates (950–1,500 mm/a) and thick porous aquifers. In contrast, the hydrogeological situation of hard rock areas is characterised by lower recharge rates due to steeper morphology and bad storage conditions in areas of high elevation (Bender et al., 2001). Difficulties for the field of water management in the area of the Czech-German border are caused by insufficient understanding of the hydrogeologic system as well as by minor interest of the responsible governmental agencies and the private institutions involved. As neighbouring regions with better groundwater resources compared to hard rock sites are present in most cases, there exist only local plans for water management. The disinterest in these Bavarian regions is increased by their geographical location in a transboundary zone. Regarding hydrogeological questions the main problems are caused by the nonconformity of national borders and natural boundaries of groundwater regions. Recently, the implementation of the EU-Water Framework Directive (WFD), associating spatial data from GIS-systems with socio-economic indicators, established new demands for catchments. To ensure a reasonable and successful water management, entire catchments have to be monitored. Therefore, the national border may not be considered as a line limiting interests of local or national authorities, as in fact all users of a transboundary catchment are responsible for water management. To simplify water related decisions and to consider the EU-Water Framework Directive, the main goal of the EU-project TRANSCAT (Integrated Water Management of Transboundary Catchments) is to create an operational and integrated comprehensive Decision Support System (DSS) for optimal water management of catchments in borderland regions.

2 TRANSBOUNDARY CATCHMENTS

The main goal of the EU-project TRANSCAT is the creation of an operational, integrated and comprehensive Decision Support System (DSS) to optimise water management in catchments of borderland regions, in context of the implementation of the EU-Water Framework Directive. Decision support systems (DSS) cover a wide variety of information systems including Geographical Information Systems (GIS). The density of data per area is controlled by the type of monitoring tool and monitoring strategy, depending on local or regional interest. For many investigated subjects, there exist several monitoring networks on different scales such as (a) state-wide monitoring networks (broad range of parameters), (b) regional networks (selected group of parameters) and c) special networks for local problems (indicating parameters or indicators) (Bender et al., 2001). The basis of such a DSS is a capacious database, containing all available data and information relevant to water management in border regions. To provide the possibility of EU- or even world wide application, the database has to be filled with data from as many countries as possible, including geological features, hydrogeological and hydrological conditions, morphology and much more. One additional step of great importance is the transformation of punctual information into spatial data. Using data from different countries within the EU implies a serious difficulty: Ensuring the comparability of data. Obviously, when data from different sources is to be combined within one database, the data may be differing concerning various characteristics such as accuracy, precision, sampling intervals, completeness, detection limits and scale.

Input meta-data for developing and verifying TRANSCAT-DSS originate from five pilot areas (PAs) across Europe. Due to the fact that a lot of problems occur in river systems with different adjacent countries, four PAs were selected to face typical questions related to contamination (agriculture, livestock farming, industrial activities, settlements) or unregulated water consumption in the headwater. The exceptional position of the fifth PA Šumava is due to the fact that only small transboundary surface water catchments exist in the area, while the extend of transboundary groundwater catchments is uncertain. The area is characterised by a predominance of protected forest regions with emerging tourism and weak economic basis. The risk of anthropogenic impacts on soil and groundwater is low.

3 PILOT AREA "ŠUMAVA"

The Czech-German catchment-cluster PA Šumava consists of Region Upper Regen (headwaters of the river Schwarzer Regen) and the Nature Park Šumava (Region Sumava, figure 1). It is drained by the rivers Vltava and Otava, on the Czech side and the river Schwarzer Regen on the German side, which belongs to the Danube stream catchment. Due to the European watershed crossing the German-Czech border region there exist only subordinate cross border surface water catchments, but transboundary groundwater catchments must be supposed. Geological units mainly consist of hard rocks, namely paragneisses and granites, which both generally show a very low permeability. Nevertheless, locally parts with increased

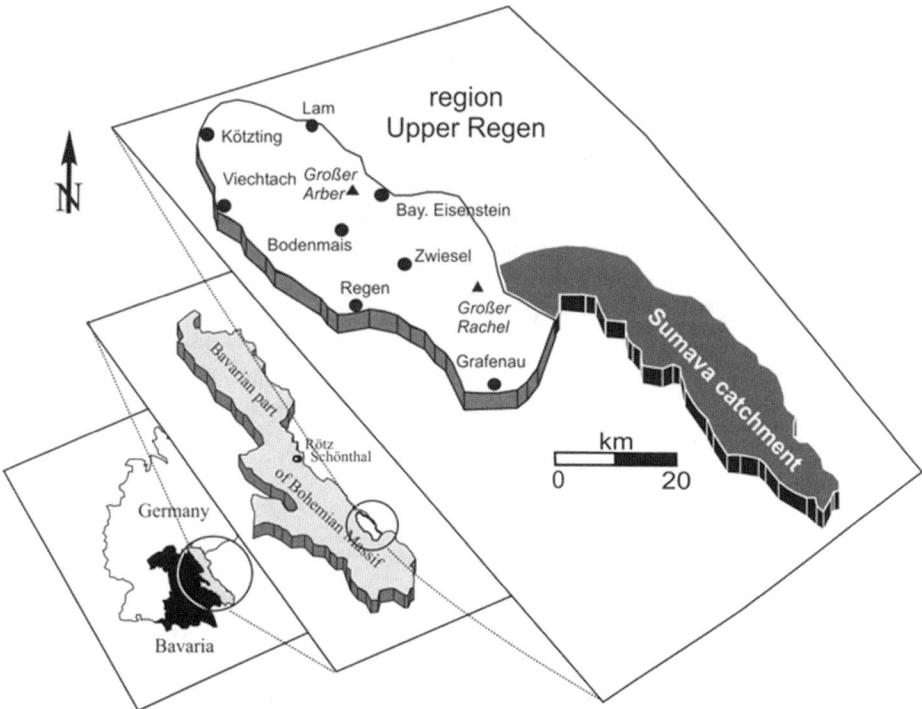

Figure 1. Location of the Pilot Area Šumava.

hydraulic conductivity can be found, such as the transition zones between lithological units, intensively stressed tectonic zones and former circulation paths. Due to these environmental conditions all management aspects have to focus on scaling effects of indicators as well as varying range of representativeness of information (Bender et al., 2004a).

The mountain range of Šumava, which is part of the Bohemian Massif, forms a historical border between the Czech Republic and Germany. Due to the political circumstances ("Iron Curtain"), the area was mainly controlled by rare military actions. Therefore the natural development of the environment is nearly uninfluenced. The forest areas in higher elevated locations are part of the "National Park Bavarian Forest" and the "National Park Šumava". In total both National Parks form the largest unified forest-region in Central Europe. On the German side, a couple of small towns and cities exist, such as Bodenmais, Zwiesel or Regen. Economically, the region largely depends on agriculture and to an increasing degree on tourism, a quite fast growing industrial sector in this area (Bender et al., 2004b). Information is mainly available from three types of measuring networks for climate data as well as for conditions of surface water and groundwater. These networks are supported and maintained by different organisations, which collect data for different parameters with different degrees of accuracy and in varying intervals, depending on their requirements. Concerning hydrochemical, morphological, hydrogeological and climatological characteristics (Vornehm et al., 2003) as well as for land use, this region can be divided into two parts: (1) highly elevated areas are predominantly characterised by forest (mainly spruce), (2) morphologically low regions (mainly on German side) are composed of farmland, forests and settlements with small industrial sites. While the higher elevated regions are more interesting for Nature Park administration than the low regions, the latter are highly important for local water management. Unfortunately, the interests of both groups are focused on different goals, so management strategies must be divided into two parts, respecting the particular spheres of interest.

On Czech side conditions are nearly the same. Due to a low density of settlements, a restrict limitation of liquid manure and inorganic fertilisers as well as the lack of pesticides (high initial costs) risks for groundwater contamination are low. After 1990 less fertile parts mostly in morphologically higher parts were abandoned or transferred into forest or grass land (Doutre 2004). Data in this area was mainly collected at test sites in the Nature Parks but without defined monitoring network and strategies. To adjust measurement programmes for a cross border network, the planned DSS can help to simplify arrangements of standardised procedures. During the first meeting of the Šumava Steering Committee a prototype of the TRANSCAT-DSS was presented to show possibilities of a bilateral data base. While water management in Nature Parks is of inferior relevance, there is an increasing interest for the morphologically low parts.

4 MAIN INDICATING PARAMETERS

In order to efficiently identify the quality of certain conditions (while this identification may take place within the planned DSS) it is necessary to define respective indicating parameters. Using questionnaires all project partners tried to find stakeholders providing their knowledge of local or regional problems, that should be solved by using a DSS. Figure 2 depicts the number and types of stakeholders, which are involved in several local steering committees, where indicators for further work are presented and discussed. These

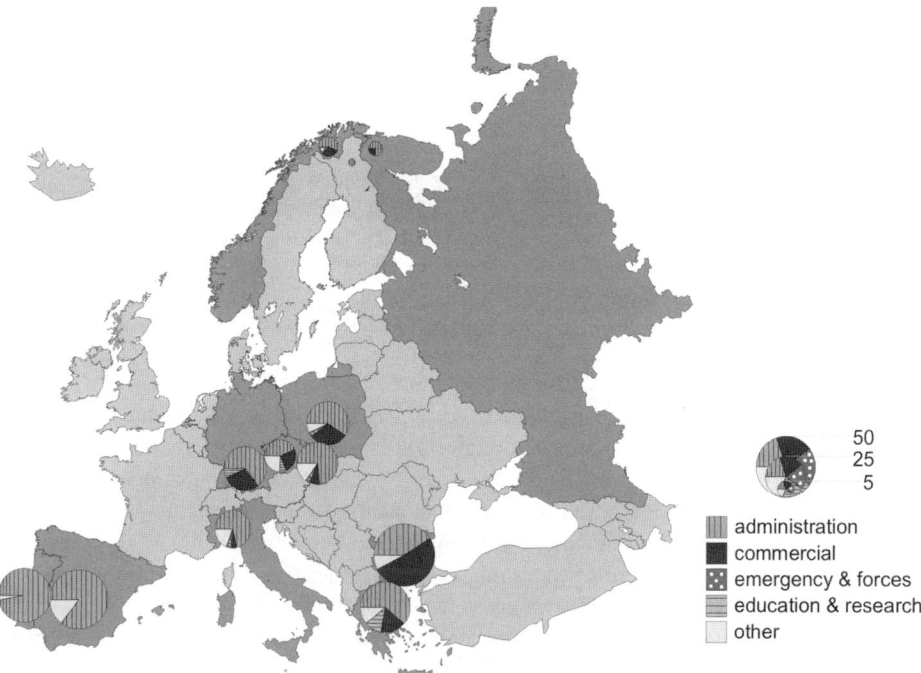

Figure 2. Overview of TRANSCAT stakeholders (Tylcer 2004).

Table 1. Selected indicating parameters for water quality (Bender et al., 2004b).

Indicator/index value	Indication
pH	– acidification (e.g. atmospheric impact from NO_x and SO_2, pyrite oxidation) – current condition of buffer system
Oxygen content	– aerobic/anaerobic conditions – oxygen consuming reactions such as decaying of organic matter
BOD	– amount of oxygen "demanded" by bacteria to break down the organic matter (such as sewage)
Specific electric conductivity	– impact of higher mineralised water
Nitrate, pestizides	– agricultural activities
Saprobic index	– existence or missing of certain species living in water (index for water quality)

indicators can translate physical and social science knowledge into manageable units of information. They can also be used in the sense of an early warning system, if the conditions described by the indicating parameters may turn into the stage of problems at a certain point in quality. This way, indicators can provide crucial guidance for decision-making. Most typically in the field of water management, these indicators refer to water quality. In this case, the respective indicators (or here rather index-values) quantify the input from certain potential sources of contamination (table 1). The type of these parameters is always connected to certain problems or tasks, what makes it highly dependent on local

Table 2. Potential indicators for the German/Czech PA.

Problem/task	Indicator
bark beetle activity (decline of spruce forest)	– nitrate – runoff
water shortage in areas supplied by groundwater	– precipitation per capita – spring discharge per capita
impact of tourism industry	– number of tourists over number of local residents, per month
acidification of soil and groundwater	– acidification index (pH, sulphate and nitrate)

and regional characteristics of the area under investigation. This means that first, the locally specific problem has to be defined, then it has to be investigated which parameter is applicable to quantify the importance/urgency of this particular problem. Some examples for the German PA Region Upper Regen are given by the following indicators (table 2). In other regions, naturally different problems and as well different indicators exist. In terms of the TRANSCAT project, indicators are expected to be somehow linked to questions, problems and tasks in the field of water management. However, as indirect importance in the field of water management may be given for various kinds of indicators which are not mandatory environmental indicators, social, economic and institutional indicators may not be disregarded. The final statement on the applicability of a certain indicator of course strongly depends on the respective data availability which is necessary to quantify the indicator.

5 WATER MANAGEMENT IN HARD ROCK REGIONS

Geological bodies are usually determined by lithological (petrographical) and stratigraphical units. In contrast spatial geometry of hydrogeological units is defined by its internal character such as distribution of porosity. Mostly lithology and stratigraphy control geometry and structure of hydrogeological units, but sometimes the hydrogeological environment is almost or entirely independent from geological features. In crystalline rocks, extension of aquifers usually does not depend on the occurence of distinct geological bodies. The effects of diagenesis, weathering and tectonic activities are more important for hard rock areas with typically shallow porous aquifers and fractured aquifers of the bedrocks. In combination with a very heterogenous geology and hydrogeological structure, it can be stated that the complex system of hydraulic and hydrogeologic interaction has not been well understood so far, making a well-planned water management hard to obtain. In comparison with extensive porous aquifers, water related work is inside a frame where heterogeneities are ubiquitous. Heterogeneities can not only be found in the lithological, hydrogeological and hydrochemical environment of the area and the respective data base, but also for inhomogeneously distributed anthropogenic impacts (Bender 2005). Besides the fact that in some parts of the area the amount of existing data is completely insufficient, another problem is imposed by data sources with and without anthropogenic influence in direct vicinity to one another, caused by the heterogeneous nature of the local geology. As the anthropogenic impact cannot always be directly identified, one is dealing with two different, incomparable data sets without even knowing about it. Such a situation may have a high influence on statistical evaluation methods for regionalisation of data (Bender 2000). In combination with

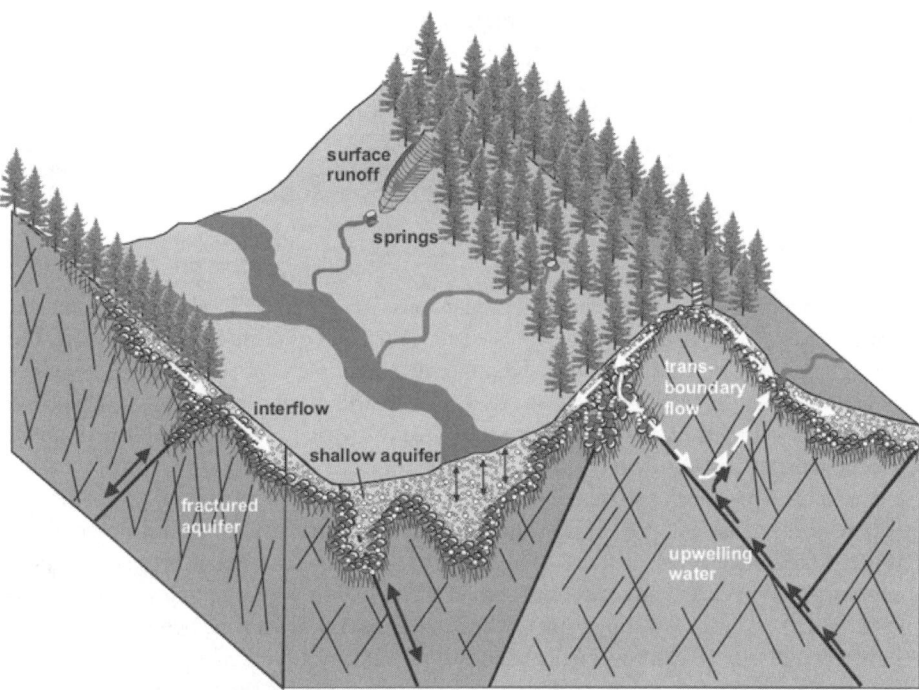

Figure 3. Scheme of the hydrogeological situation in transboundary hard rock regions (Bender 2005).

a small data base, conclusions for these parameters are afflicted with high uncertainties. The hydrogeological situation of PA Šumava is of great interest and more complex than in other TRANSCAT-PAs. It is characterised by a combination of shallow porous weathered materials and fractured hard rock aquifers of the basement beneath. The most important for the hydraulic conditions is therefore the degree and kind of weathering and fracturing. The weathering process and the presence of generally better permeable Quaternary deposits result in an increase of groundwater storage and faster flow. Transmissivity variations of samples containing Quaternary deposits are usually lower than that of samples of hard rocks without coverage (Krásný 1998). This indicates an equalising effect of hydraulically more homogeneous deposits. Within the subterranean catchment, transboundary groundwater fluxes most probably occur (figure 3). Detailed investigations in the Upper Palatinate Forest show a broad variability of thickness of covering layers, which ranges from 0 up to 100 m.

Spring systems can be influenced by upwelling groundwater from deeper aquifers (Bender 2000, Breuer 1997). Unfortunately physicochemical information are predominantly available for shallow aquifers, where available data mainly originate from numerous springs. Due to the high number of springs mostly located in the highly elevated parts of the area, there was no mandatory need to drill wells for local water supplies. Therefore, geological and hydrogeological knowledge of local hydraulic conditions is extremely weak (Krásný 1996). The upper areas with elevations over 1,000 m a.s.l. are part of the Nature Parks Bavarian Forest and Šumava, meaning good protection of groundwater due to limitations of mostly all anthropogenic activities. One of the main problems in this mountainous region is acidification of soil and groundwater due to low buffer capacities of soils and gruss layers

(Hrkal *and* Fottová 1999). As a result of bark beetle activities or lumbering in combination with spruce monocultures and with input of atmospheric deposition (SO_2, NO_x) groundwater gets more acidic, enhancing the mobility of heavy metals and aluminum. Creation of pollution load maps using risk analysis methods of Hrkal (2001) point to the high vulnerability of morphologically high parts (Vornehm *et al.*, 2003).

6 HETEROGENEITIES

Hydrogeological data sets are never homogeneous. The spatial distribution of physicochemical or hydraulic parameters depends on the extension of the area under investigation and on its relation to the size of decisive inhomogeneity elements of the respective hard rock environment (Krásný 1998). On a local scale distribution of parameters is chaotic as a result of prevailing systems of fractures and fissures. Variations are mainly attributed to distinct character such as abundance of faults and joints or changing thickness of weathering zone and covering layers (Bender 2000, Breuer 1997). In larger areas (sub-regional or medium scale) mean and prevailing values are mostly similar. Additional differences are caused by inhomogeneous elements of higher order. In case of hard rock areas it is important to distinguish at least three different compartments: (1) porous aquifer system, (2) fractured aquifer system and (3) unweathered bedrock. The transition zone between both types of aquifers is discussed controversially (Raum 2002, Krásný 1996, Larsson 1987, Saker and Jodan 1977). Figure 4 shows the different existing concepts for subdivision of the underground system, while the geological cross-section in figure 5 gives an example for heterogeneities in a typical crystalline hardrock environment as the Bavarian Forest.

Regional tendencies due to geological or petrographical similarities or neotectonic activities are responsible for heterogeneities on a regional scale (Rohr-Torp 1994, Havlík and

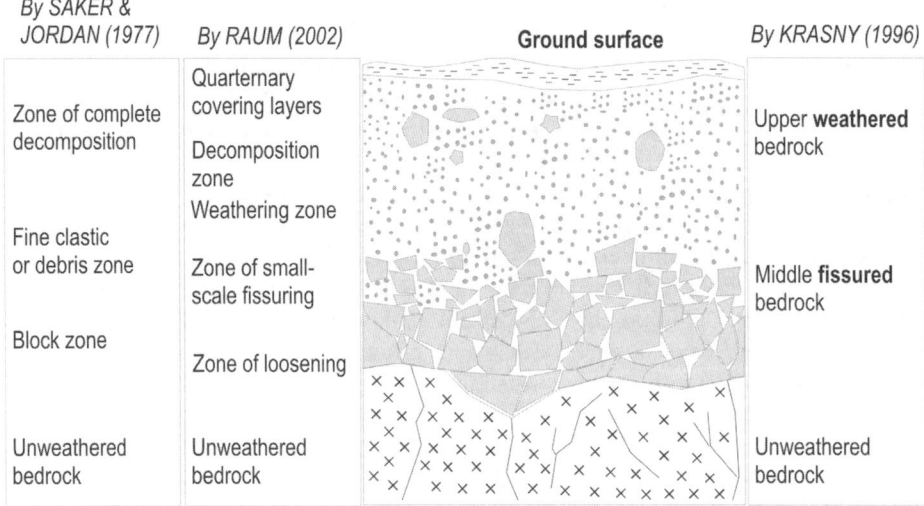

Figure 4. Conceptual subdivision of crystalline bedrock and its weathering products (Rubbert *et al.*, 2005).

Krásný 1998). The degree of fracturing in the Bohemian Massif is a result of tectonic activities during Neogene and Quaternary, whereas these effects are very similar and can be well compared to isostatic uplift after Quaternary Fennoscandian glaciation. Combination of different scales shows that the hydrogeological situation of hard rock areas should not be considered as regionally homogeneous but rather as a complex system where hydraulic parameters follow morphological or tectonical features. Compared to hydrogeological information it is often easier to obtain hydrochemical data. Unfortunately, existing information can not always be compared due to differences of (a) sampling methods, (b) determination methods, (c) seasonal impacts or (d) different scales of data. For the integration of all available information it is necessary to use different scales to depict detailed information, local mean values, regional mean situation or supraregional conditions. Combining data from different scales results in an inconsistent data set with incompatible ranges of prediction and weighting factors (Bender *et al.*, 2004a). Due to the fact that punctual hydrochemical information as well as other hydrogeological data has to be transferred into spatial distribution maps, natural spatial heterogeneities may cause further detrimental impacts on the results. Detailed investigations on a regional and local scale showed different types of difficulties concerning the regionalisation process. With respect to data density the most important factor is not the number of information but the representative range of available information, similar to the representative elementary volume for integral measurements (Mieseler and Wisotzky 2005). Furthermore the monitoring network configuration has an extreme impact on regionalised results due to more or less ideal spatial mathematical correlation of information (Bender *et al.*, 2002). Finally, combination

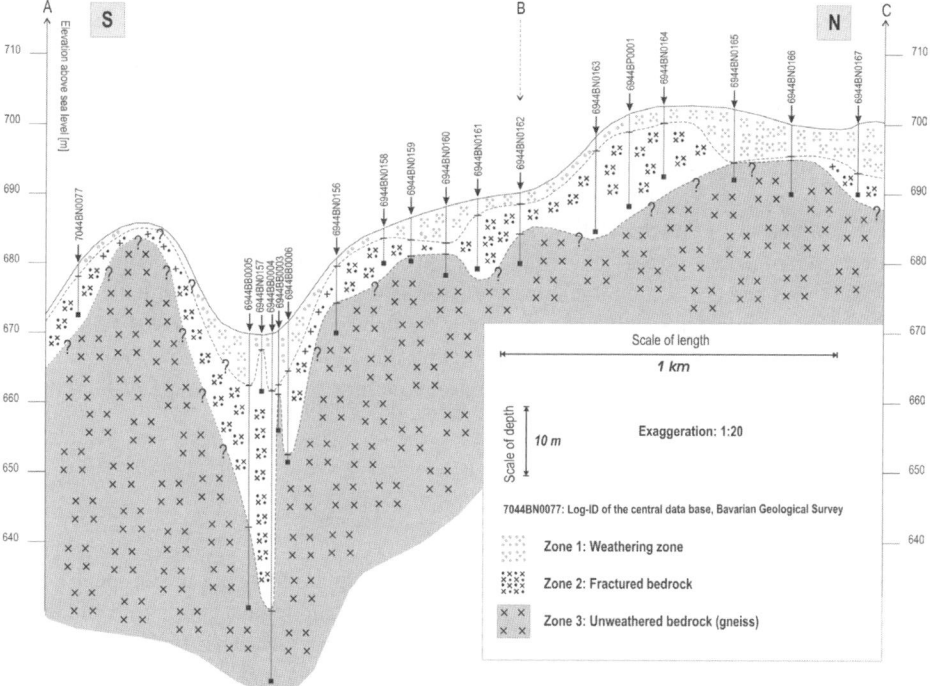

Figure 5.   Geological cross section near the city of Regen, Bavarian Forest (Rubbert *et al.*, 2005).

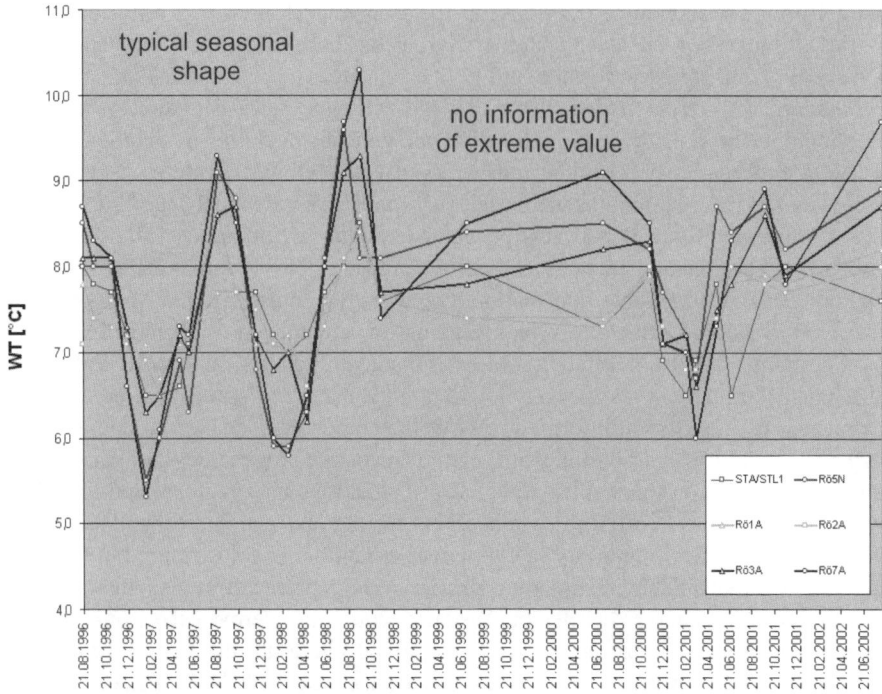

Figure 6. Impact of sampling interval on data sets (Bender et al., 2004b).

of different data sets with different kinds of impacts results in smoothing effects and high uncertainties due to used interpolation methods, particularly in the vicinity of extreme values deviating significantly from the geogenic background (Bender 2000).

A sampling interval is nothing else than a temporal aspect of scaling effects (figure 6). There are two aspects determining sampling intervals: (1) costs and effort of data acquisition, storage, evaluation and interpretation, and (2) quality of data. Large intervals reduce costs but for most parameters upper limits of the interval exist to enable correct interpretation. To minimise effort and costs, the largest interval still suitable to obtain data which is detailed enough to get the information that was aimed for has to be chosen. If seasonal or daily trends have to be monitored, then the interval conforms with the scale of question. The selection of a larger interval results in a missing of extreme values so that the shape of the curves cannot be regarded as significant.

7 CONCLUSION

Even though a basic level of conformity is given for the legal framework in European countries, there are still many particularities concerning norms and regulations applicable in the different countries. Furthermore sampling and analytical strategies can differ between companies or laboratories. Due to a combination of different methodologies, network density and configuration as well as monitoring strategies, transboundary data pools are mostly heterogenous or – in other words – consist of several homogenous data sets. Main task for a transboundary water management is the development of a multi-scale, integrated flexible DSS, which consists of several modules for various topics such as climatic, environmental or socio-economic processes. Basic principle of work is a common standard, which can be used for data acquisition. Such a standardized method will definitely be an enormous progress for future applications. Working in a test site which is predominantly characterised by extremely heterogeneous natural conditions as well as a heterogeneous data base of existing information, the field of "data acquisition" is highly important. First of all, for an optimised usage of the DSS, the co-operation of countries on both sides of the border is essential. To account for existing problems, communication between local steering group members including representatives from governmental authorities, municipalities and local water suppliers must be promoted. Regarding water management in hard rock areas, it is necessary to find evaluation methods enabling predictions of the hydraulic and hydrogeologic interactions in this complex system. A more detailed data analysis is recommended to continue to reveal all the complexity of the environment. A complicated hierarchic system of inhomogeneity elements of different size should be taken into account by implementing regional conceptual and numerical models. As a conclusion of data evaluation, for areas with high contamination risk only detailed investigations are useful to detect and visualize changes at a large scale. Using a higher scale hierarchy results in a mixing of two types of data, namely in this particular case two types of water resulting in mean or buffered values of low significance. Starting points for an approach are GIS-based calculations using available spatial data in combination with weighted levels (Hrkal et al., 2003) or anthropogenic indicators which can be used as tracers to understand hydraulic processes.

ACKNOWLEDGEMENT

This work is part of the TRANSCAT project on "Integrated water management of transboundary catchments" realised in the frame of the contract EVK1-CT2002-00124 of the 5th Framework program (Energy, Environment and Sustainable Development).

REFERENCES

Bender S (2005) Die Aussageunschärfe bei der Verwendung heterogener Datensätze im Rahmen wasserwirtschaftlicher Fragestellungen. Bochumer Geowiss. Hefte (in Vorb.)

Bender S (2000) Klassifikation und genetische Entwicklung der Grundwässer im Kristallin der Ober-pfalz/Bayern. Münchner Geol. Hefte B10

Bender S, Mieseler T, Rubbert T, Wohnlich S (2004a) Scaling Effects in a multi-layered DSS-structure. In: Proc. of Conf. Integrated Water Management of Transboundary Catchments, S03–07

Bender S, Mieseler T, Rubbert T, Wohnlich S (2004b) Collection of monitoring strategies effective in countries participating in the TRANSCAT project. TRANSCAT-report, WP4-DL4.1-1

Bender S, Einsiedl F, Wohnlich S (2001) Scheme for development of monitoring networks for springs in Bavaria, Germany. Hydrogeology Journal 9, 2: 208–216

Bender S, Nickol P, Wohnlich S, Zöllner R, Zoßeder K (2002) Detection of contaminant plumes below ensemble-protected urban areas – A case study for the gravel plain of Munich. Proc. European Conference on Natural Attenuation, 218–219

Breuer B (1997) Hydrogeologische Gegebenheiten in der Verwitterungszone im Umfeld der Kontinentalen Tiefbohrung (KTB), Oberpfalz. PhD, Univ. Erlangen-Nürnberg

Doutre J (2004) Test Site Šumava – Czech Side. In: Bender S, Mieseler T, Rubbert T, Wohnlich S (eds.) Collection of monitoring strategies effective in countries participating in the TRANSCAT project. TRANSCAT-report, WP4-DL4.1-1, pp 124–130

Havlík M, Krásný J (1998) Transmissivity Distribution in Southern Part of the Bohemian Massif: Regional Trends and Local Anomalies. – In: Annau R, Bender S, Wohnlich S (eds.): Hardrock Hydrogeology of the Bohemian Massif, Münchner Geol. Hefte B8, 11–18

Hrkal Z (2001) Vulnerability of groundwater to acid deposition, Jizerské Mountains, northern Czech Republic: construction and reliability of a GIS-based vulnerability map. Hydrogeology Journal 9: 348–357.

Hrkal Z, Fottová D (1999) Impact of athmospheric deposition on the groundwater quality of the Czech Republic. Hydrogéologie 2: 39–45

Hrkal Z, Bender S, Sanchez Navarro JA, Martin C, Vayssade B (2003) Landscape-use Optimisation with Regards to Groundwater Protection Resources in Hard Rock Mountain Areas (LOWR-GREP); some results from an European research program (5th PCRD). In: Proc. 5th Int. Symp. of Water, 335–343

Krásný J (1998) Groundwater discharge zones: sensitive areas of surface-water – groundwater interaction. – In: Van Brahana J, Eckstein Y, Ongley LK, Schneider R, Moore JE (eds.): Proc. Gambling with groundwater, pp 111–116

Krásný J (1996) Hydrogeological environment in hard rocks: an attempt at its schematizing and terminological consideration. In: Krásný J, Mls J [eds.]: Hardrock hydrogeology of the Bohemain Massif, Acta Univ. Carolinae Geologica, 40, 2: 123–133

Larsson I [ed.] (1987) Les eaux souterraines des roches dures du socle. Études et rapports d'hydrologie, 33

Mieseler T, Wisotzky F (2005) Flächendifferenzierte Bestimmung des Nitrateintrages in das Grundwasser mit Hilfe der neu entwickelten Nitratelutionsmethode. Grundwasser (in Vorb.)

Raum KD (2002) Markierungstechnische, bruchtektonisch-gefügekundliche und fotogeologische Untersuchungen zur Ermittlung der Grundwasserfließverhältnisse in der Verwit-terungszone kristalliner Gesteine in Quellgebieten des Oberpfälzer/Bayerischen Waldes (Ost-Bayern/Deutschland). PhD Univ. Erlangen-Nürnberg

Rohr-Torp E (1994) Present uplift rates and groundwater potential in Norwegian hard rocks. Geological Survey of Norway, Bulletin 426: 47–52

Rubbert T, Mieseler T, Bender S (2005) Hydrogeological modelling in the combined porous-fractured aquifer system of the Bavarian Forest. In: Proceedings of the 2nd Workshop on Hardrock Hydrogeology (IAH Commission on Hardrock Hydrogeology Iberian Regional Working Group), Evora, Portugal, 18.-21.05.2005, 9 S

Saker I, Jordan H (1977) Zu hydrogeologischen Eigenschaften der Verwitterungszonen erzgebirgischer Gneise. Zeitschrift für angewandte Geologie, 23, H 12, 606–611

Tylcer O (ed.) (2004) Specifications of requirements for the DSS system design. TRANSCAT-report, DL3.2

Vornehm Ch, Bender S, Wohnlich S (2003) Geochemical zoning of soil and groundwater due to atmospheric deposition in the "Arber" region, Bavarian Forest, South Germany. In: Proc. Diffuse Input of Chemicals into soil and groundwater – Assessment and Management, pp 161–170

CHAPTER 3

Combined use of indicators to evaluate waste-water contamination to local flow systems in semi-arid regions: San Luis Potosi, Mexico

A. Cardona[1], J.J. Carrillo-Rivera[2], G.J. Castro-Larragoitia[1] and E.H. Graniel-Castro[3]

[1]*Facultad de Ingeniería-UASLP, Av. Manuel Nava 8, Zona Universitaria, San Luis Potosí Cd, México*
[2]*Instituto de Geografía-UNAM, CU, Delegación Coyoacán, México D. F.*
[3] *Facultad de Ingeniería, Universidad Autónoma de Yucatán, Cordemex, Mérida, Yucatán.*

ABSTRACT: Geochemical and microbiological data were collected for shallow groundwater flow systems of a semi-arid area of Central Mexico to investigate the effects of contamination from different sources. Nitrate and nitrite concentrations, and faecal coliform counts exceeding the Mexican Health Agency's maximum contaminant level were determined in most samples within the study area, suggesting significant impact on local groundwater flow systems from anthropogenic activities. Values of alkalinity (up to $600\,mgL^{-1}$ as $CaCO_3$), Cl (up to $400\,mgL^{-1}$) and SO_4 (up to $1,030\,mgL^{-1}$) are quite high in some places, especially near the industrial park where waste-water injection to shallow depths used to be a common practice. Reported hydraulic head conditions show that a strong vertical gradient exists between the shallow contaminated zones and deeper regions. However, direct transfer of contaminated water is also observed via poorly constructed or abandoned boreholes, down which shallow contaminated water can flow, polluting deep sources of potable water supply. This condition constitutes one of the greatest threats to potable water sources in the area.

Keywords: waste-water, contamination, groundwater flow systems, hydrochemistry, San Luis Potosi, Mexico.

1 INTRODUCTION

In most semi-arid zones of Mexico groundwater is the main source for potable, agricultural and industrial use. At present, the study area, San Luis Potosi (SLP) City (figure 1) is one of the conurbations of Mexico with the highest annual growth rate ($\approx$5–7%) with about 9,00,000 inhabitants. In the last 30 years, some 95% of its total water supply for human consumption has been obtained from groundwater. Prevailing semi-arid conditions make natural water resources limited, such that waste-water reuse for irrigation purposes makes a lot of sense. However, irrigation with untreated waste-water can cause deterioration of shallow groundwater quality, further limiting available water resources for potable use. Shallow

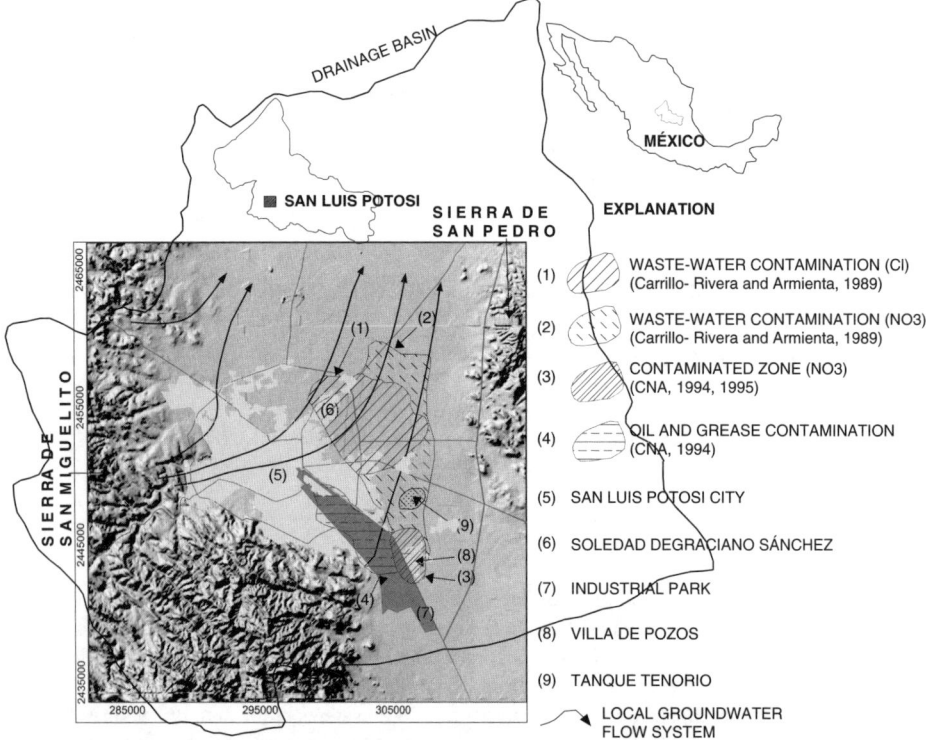

Figure 1. Location of study area showing where shallow groundwater contaminated with waste-water have been identified in previous investigations. General shallow groundwater flow lines in the horizontal plane are also shown.

groundwater was used for drinking water purposes in the study area until some 100 years ago. As the drainage basin is naturally closed, waste-water produced by inhabitants has drained to the parts with lowest elevation and has usually been consumed in the nearby district of Soledad de Graciano Sanchez (SGS) for crop irrigation. In 1995 an estimated discharge of about $1.9\,m^3s^{-1}$ of waste-water (95% urban domestic and commercial and 5% urban industrial) was used for the irrigation of over 2200 ha (CNA, 1995). In 2001 the city obtained its groundwater supply (about $3.0\,m^3s^{-1}$) from more than 120 boreholes tapping a deep aquifer located below the shallow aquifer.

An aquifer system has been defined in the study area; it consists of a shallow unconfined aquifer and a deep aquifer (Carrillo-Rivera et al., 1996). Most of the boreholes tapping the deep aquifer (65% in volume) are used for drinking water supply; boreholes that are located within the conurbation boundaries, situated broadly within the periphery of the highway ring shown in figure 1. Currently, there is a lack of evaluation of the diffuse contamination affecting shallow groundwater by waste-water disposal practices as well as studies of shallow groundwater quality for the SLP area. The reason for this could be that deep groundwater has been the major source of water for potable supply for the last 60 years (the first deep boreholes were drilled in the 1940's). Stretta and Del Arenal (1960) noticed contamination of shallow groundwater in the zone where raw waste-water was used for irrigation purposes, but no groundwater quality information was presented to support this conclusion. The first study specifically investigating shallow groundwater quality was presented by Carrillo-Rivera and Armienta (1989). They delineated a zone east of SLP City where shallow groundwater was contaminated with urban waste-water, producing concentrations of NO_3, Cl and HCO_3 well above natural baseline conditions (figure 1). Later CNA (1994, 1995) and Geoingenieria Internacional (1996) studied with some detail the industrial park in the portion adjacent to Villa de Pozos, detecting high concentrations of several organic compounds, NO_3 and some heavy metals. These data suggest that historical and present waste-water disposal is likely to have caused significant water quality deterioration in shallow groundwater, with waste-water infiltration modifying the initial baseline water chemistry.

Considering that urban waste-water disposal practice typically excludes secondary and tertiary treatment stages in most cities of the country, this environmental problem is considered to be a common issue in cities with more than 1,50,000 inhabitants. Yet it has not been widely documented. Some examples of groundwater contamination due to poor waste-water management have been described in Mexico for different agricultural areas that use raw waste-water for irrigation. For example in the Mezquital valley (Hidalgo state) and Leon plain in Guanajuato (Chilton et al., 1996), also in the Hermosillo coastal aquifer (Steinich et al., 1998; Silva-Lugo, 2005), in the Tecamachalco region in Puebla (Dominguez-Mariani et al., 2004) and in the Comarca Lagunera in Coahuila (Molina-Maldonado et al., 2001). In these regions the environmental impact on groundwater was typified by high NO_3 and Cl concentrations, as well as total and faecal coliform positives and occasionally anomalous heavy metals concentration, all of which have produced deterioration in groundwater quality.

Understanding the nature and scale of the contamination problems actually present in shallow groundwater of SLP City helps provide an insight into the processes involved and could be used as a tool to propose mitigation programs. In this paper, the combined use of several indicators of contamination such as chemical (inorganic) major elements, nutrients, organic pollution load and biological information helps identify current shallow groundwater quality deterioration in the study area, and also defines the extent that waste-water usage has changed the initial (pre-development) natural baseline conditions. Multiple indicators help gain a more complete assessment of the contamination of shallow groundwater affected by poor waste-water disposal practices.

2 REGIONAL SETTING

The drainage basin of SLP is one of the several closed basins existing in the north-central part of Mexico. The basin has a surface drainage of about $1,900 \, km^2$; however, the actual area covered in this research study (figure 1) considers just the conurbation. The abrupt relief of the surrounding mountains of Sierra de San Miguelito (SSM) to the west, and Sierra de San Pedro, to the east, is composed of acid extrusive volcanic and calcareous rocks, respectively. These sierras have an elevation in excess of $2,300 \, m \cdot amsl$ and slope towards the plain of the drainage basin which has an altitude of about $1,900 \, m \cdot amsl$. The climate is semi-arid, potential evaporation ($\approx 2,000 \, mm \, annum^{-1}$) significantly exceeding mean annual precipitation ($\approx 400 \, mm$), and the rainy season occurs mainly between May and October. The mean annual air temperature is around $17.5° \, C$, while the summer mean temperature is around $21° \, C$.

2.1 Geology

The closed drainage basin of SLP occupies a graben structure developed during the Oligocene. A thick (>1,500 m) sequence of extrusive volcanics (Tertiary ignimbrites, lava flows and tuffs) and alluvial materials were deposited, covering Cretaceous limestone and calcareous mudstone outcropping in folded NW-SE structures in the Sierra de San Pedro. The volcanic rocks have been differentiated into several formations (Labarthe-Hernández *et al.*, 1982); most of them are felsic to intermediate in nature (ranging from rhyolite to latite) but a minor component of mafic (andesite) rocks is also present. These rocks are affected by different regional faulting systems (NE-SW and NW-SE). A clastic sequence (gravel, sand, silt and clay derived from the surrounding Tertiary volcanic and Cretaceous calcareous rocks) was deposited on top of the volcanics as basin fill material in the graben structure and is up to 500 m thick. Borehole logging data indicate the presence, throughout the basin, except at the edges, of a 50–150 m thick bed of fine grained compact quartzitic sand fully enclosed within the clastic sequence.

2.2 Hydrogeological features

The hydrogeology of the drainage basin of SLP has been defined in detail by Carrillo-Rivera (1992) and Carrillo-Rivera *et al.* (1996). In this closed drainage basin, two main hydrogeological units (named locally shallow and deep aquifers) are separated in the vertical direction by the compact sand layer, which has a low horizontal hydraulic conductivity. The shallow aquifer studied in detail in this investigation is in alluvium, and under water-table conditions, perched on the compact sand layer. Boreholes tapping this aquifer unit are around 5–100 m deep, with a typical average extraction rate of 0.005–$0.010 \, m^3 s^{-1}$. The depth to water-table is in the order of 5–35 m, with lower values observed in the southwest part of the plain and higher values towards the east. Measured elevation of the water-table in shallow boreholes illustrates a general flow direction in the horizontal plane from southwest to northeast (figure 1). Horizontal hydraulic gradient is about $0.02 \, mm^{-1}$ for the southwest region and 0.006–$0.008 \, mm^{-1}$ for the eastern zone. However, vertical hydraulic gradients are higher than the horizontal ones in most of the area; a vertical hydraulic gradient value of about $1 \, mm^{-1}$ has been reported in the southern part of SLP City (Geoingenieria Internacional, 1996). Pumping-test analyses reported by Carrillo-Rivera (1992) indicate average horizontal hydraulic conductivity for the shallow aquifer in the SGS zone to be of about $2 \times 10^{-4} \, ms^{-1}$. The average horizontal hydraulic conductivity of the compact sand layer was computed to be $\approx 10^{-9} \, ms^{-1}$, as a statistical mode from 26 point-piezometer slug-tests using the Hvorslev (1951) method. The slug-tests were carried out along the upper 20 m of the compact sand layer, in the southern part of SLP City (Geoingenieria Internacional, 1996). A complete description of the deep aquifer is presented elsewhere (Carrillo-Rivera, 1992; Carrillo-Rivera *et al.*, 1996; 2002).

2.3 Groundwater flow system

From a regional point of view, recharge under natural conditions is supposed to occur if favourable meteorological conditions are present along zones in the fractured uplands to the west of the study area, and along ephemeral streams after main storm precipitation events leading to runoff generated in the SSM. No major discharge zones have been identified either at present time or in the past, low yield ($<0.001\ m^3s^{-1}$) ephemeral springs and seepage occur only in the south-western part of the study area, adjacent to SSM. Local, intermediate and regional groundwater flow systems, as conceptually described by Tóth (1995), have been identified in the region. The former is related to the shallow aquifer and the others to the deep aquifer (Carrillo-Rivera et al., 1996). Whilst the local system is investigated in this study, information regarding the intermediate and regional systems is presented as a reference. The intermediate flow system has recharge generated within the east and south-south-east portion of the basin, as runoff infiltrates in the piedmont area and, in places, on the plane of the basin; this groundwater circulates at shallow (200–400 m) depth in the granular basin fill material. The regional flow system is associated with water circulating along the fractured volcanic units that outcrop in the SSM and elsewhere. Water extraction from the deep aquifer is significant; the regional system alone supplies about 70% of the total volume that is annually used for SLP City. In addition, due to the steep vertical hydraulic gradient between shallow and deep aquifers, some water is infiltrating from the former to the latter, representing a third component of groundwater abstracted from the deep aquifer.

3 METHODS, SAMPLING AND ANALYSES

The present shallow groundwater quality associated with local flow systems was monitored via discharge of shallow hand-dug wells and boreholes. Additionally, some urban and industrial waste-water discharges were also monitored as well as the main urban and industrial waste-water reservoir in the area (Tanque Tenorio), which is located to the east of the SLP-City with an open area of 210 ha (figure 1). Water samples were collected within the body of Tanque Tenorio and in some sewage mains discharging industrial and urban waste-water. The chemical composition of precipitation was defined and compared with the quality of the local groundwater flow system. Rainfall samples were collected during individual storm events throughout the rainy season (July–November) during a one-year period. Rainwater was filtered (0.45 μm) and collected in Nalgene™ bottles just after the storm event to minimise evaporation and were analysed for Cl and SO_4. The groundwater samples were obtained from a selection of the more than 300 private hand-dug wells and boreholes tapping the shallow aquifer; some were selected according to the location of known waste-water infiltration structures, both designed systems (land application of urban waste-water and injection boreholes disposing industrial waste-water) and unplanned releases (leakage from surface impoundments). Shallow boreholes were selected for water sampling according to the location of known contamination sources, most of them being used for irrigation purposes. They are about 1–3 m in diameter, 5–30 m deep with a saturated thickness of about 2–10 m below the water-table. On-site field measured parameters included water temperature, pH, dissolved oxygen (DO), redox potential (Eh, Pt-electrode), and specific electrical conductivity (SEC). Portable meters with probes encased in a closed flow-through isolation cell to ensure the exclusion of atmospheric gases and improve measurement stability were used at all times. Calibration for pH determination was made at every site using 7.0 and 4.0 buffers and allowing time for water and electrode temperature equilibration. Stabilization of readings for pH and redox potential was achieved between 10–20 minutes after measurements started; readings were taken after steady conditions were observed.

Groundwater and waste-water samples for major ions and nutrients were collected and preserved according with well established quality assurance protocols. One field filtered sample (0.45 μm) was taken at each site in a double acid-washed, well rinsed, low density polyethylene bottle and acidified with high purity HNO_3 so as to lower the pH of the sample below 2 to assure metals stabilization. Other sample for major element analyses was taken unfiltered without acidification. Samples for NO_3 and NO_2 determinations were acidified to pH of 2 with H_2SO_4. Additional samples for microorganisms were collected in sterile plastic bags. Samples for major elements, nutrients and microorganisms were stored at 4°C during sampling procedures, and transported to the laboratory and analysed on a daily basis. Mexican Standard procedures were utilized in the collection and preservation for microorganisms (total and faecal coliforms), chemical oxygen demand (COD) and oil and grease (OG).

Groundwater samples for were collected in two sampling periods. The first sampling included 44 shallow hand-dug wells, distributed over the plane of the basin, and was performed during the rainy season. The second sampling was carried out at the end of the following dry season in 33 selected sites, where contamination was detected with the analytical results of the first sampling. During the second sampling, waste-water samples were also collected in different sectors of Tanque Tenorio.

Analytical laboratory determinations we carried out using Mexican Standards Methods (NMX). Alkalinity was measured using volumetric titration with bromocresol green-methyl red indicator (NMX-AA-036-SCFI-2001). Chloride was determined by argentometric method (NMX-AA-073-SCFI-2001). Sulphate was determined by turbidimetric method using barium chloride (NMX-AA-074-SCFI-2001). Nitrate and NO_2 concentrations were analysed with automated colorimetry (NMX-AA-079-SCFI-2001). Microorganisms were determinated by the Method of Filtration in Membrane (NMX-AA-102-1987). The concentration values of PO_4 were determined using the stannous chloride method (NMX-AA-029-SCFI-2001). Sodium and K were determined by flame photometry, and total hardness by volumetric titration with H_2SO_4 and EDTA indicator (NOM-AA-51-1981). Chemical Oxygen Demand was analysed with the NMX-AA-030-SCFI-2001 method and OG by extraction with hexane (NMX-AA-005-SCFI-2000).

4 RESULTS AND DISCUSSION

4.1 Water chemistry of major constituents

Summary statistics for field-collected parameters, major ion constituents, nutrients, microorganisms, OG and COD are given in table 1. Cases elsewhere show that shallow groundwater temperature is generally similar to the annual mean air temperature, however in the study area mean measured values were similar to local mean summer air temperature, suggesting recharge is taking place either during the rainy season and infiltration of waste-water occurs under the city from septic tanks or sewer leaks. Dissolved oxygen in shallow groundwater is generally present, redox potential values are above 0.35 V, suggesting aerobic and oxidizing conditions for the local groundwater flow system, which is consistent with an observed general lack of natural organic matter in the basin fill sediments. The distribution of values of SEC in the horizontal plane shows a clear regional trend that agrees with the general groundwater flow direction, values varied between

Table 1. Summary statistical data for field-collected parameters, major elements, nutrients, microorganisms, oil and grease and chemical oxygen demand in the San Luis Potosi shallow groundwaters.

	Field parameters					Major elements (mgL^{-1})								Nutrients (mgL^{-1})			Micro-organisms CFU/100 ml			
	T°C	pH	Eh	SEC	DO	Ca	Mg	Na	K	HCO$_3$	Cl	SO$_4$	NO$_3$	NO$_2$	PO$_4$	CT	CF	OG	COD
First sampling																			
Maximum	27.2	11.47	0.473	2995	5.9	384.0	45.7	165.0	119.0	526.6	403.7	1034.9	182.6	3.58	0.80	1.95E+05	5.65E+03	4.19	87.27
Minimum	18.1	5.29	0.313	158	0.8	8.0	1.0	12.1	10.5	12.6	4.7	20.6	0.4	0.00	0.01	1.80E+01	0.00E+00	0.29	1.00
Average	21.6	6.76	0.345	1036	3.2	95.2	15.1	65.6	39.0	239.6	79.3	153.6	47.7	0.15	0.09	1.97E+04	7.27E+02	1.20	9.52
Median	21.4	6.66	0.339	910	3.2	69.3	13.6	51.2	34.2	239.5	48.1	113.3	34.1	0.00	0.01	2.85E+03	8.98E+01	0.50	1.00
Est. Dev.	1.8	0.79	0.028	695	1.4	83.7	10.5	43.1	23.0	139.3	85.4	158.7	47.7	0.55	0.16	4.84E+04	1.50E+03	1.16	16.31
75 Percentile	22.4	6.96	0.346	1307	4.2	124.8	20.4	82.7	49.1	326.8	108.5	158.8	68.4	0.03	0.12	8.98E+03	3.95E+02	1.44	13.60
25 Percentile	20.7	6.45	0.334	495	2.0	31.3	6.8	36.8	25.0	119.7	20.9	69.9	9.1	0.00	0.01	7.05E+02	3.25E+00	0.50	1.00
n	55	55	32	55	55	55	55	55	55	55	55	55	55	55	55	16	16	16	50
Second sampling																			
Maximum	23	8.08	0.388	3190	7.2	598.4	48.9	278.0	122.5	554.1	434.9	1534.5	1917.6	9.17	3.86	1.52E+05	1.20E+04	8.47	96.00
Minimum	16.3	5.91	0.251	615	0.0	50.5	4.6	11.5	16.5	92.2	4.9	69.4	1.6	0.00	0.09	1.00E+01	0.00E+00	0.50	1.00
Average	20.1	7.02	0.347	1359	3.2	169.2	26.3	115.1	43.6	295.5	131.0	229.7	150.4	0.51	0.63	1.20E+04	1.18E+03	1.69	22.76
Median	20.1	6.99	0.350	1246	3.4	129.7	27.7	82.5	36.5	255.2	119.9	156.1	49.3	0.01	0.31	3.50E+03	2.65E+02	0.50	16.00
Est. Dev.	1.6	0.49	0.026	610	2.0	115.5	11.4	75.6	23.4	119.6	102.6	265.9	336.0	1.85	0.88	2.70E+04	2.27E+03	1.90	24.79
75 Percentile	21.5	7.26	0.362	1803	4.6	187.6	35.2	145.0	49.5	339.2	168.7	236.0	132.1	0.06	0.31	7.95E+03	1.30E+03	2.98	30.00
25 Percentile	19.5	6.74	0.337	824	1.5	102.3	16.8	56.4	32.0	210.7	55.8	95.8	31.3	0.00	0.31	1.86E+03	4.00E+01	0.50	1.00
n	33	33	32	33	32	32	32	33	33	33	33	33	33	33	33	33	33	33	33

(*Continued*)

Table 1. (continued)

	Field parameters					Major elements (mg L^{-1})								Nutrients (mg L^{-1})			Micro-organisms CFU/100 ml			
	T°C	pH	Eh	SEC	DO	Ca	Mg	Na	K	HCO$_3$	Cl	SO$_4$	NO$_3$	NO$_2$	PO$_4$	CT	CF	OG	COD

Tanque Tenorio waste-water

	T°C	pH	Eh	SEC	DO	Ca	Mg	Na	K	HCO$_3$	Cl	SO$_4$	NO$_3$	NO$_2$	PO$_4$	CT	CF	OG	COD
Maximum	28.8	8.22	0.275	1940	0.2	179.6	44.0	312.5	42.5	1058.5	140.0	147.8	3.9	3.22	25.45	3.55E+08	7.70E+07	38.04	1028.16
Minimum	23	7.34	0.058	1476	0.0	72.5	34.8	222.5	25.5	739.4	102.5	61.5	1.6	0.00	6.99	4.27E+06	4.95E+02	9.09	299.88
Average	26.3	7.75	0.183	1610	0.1	109.3	37.8	256.9	30.5	845.2	125.0	103.8	2.9	1.61	17.39	1.80E+08	2.98E+07	19.07	593.73
Median	26.6	7.73	0.215	1513	0.0	92.6	36.2	246.3	27.0	791.5	128.7	102.9	3.0	1.61	18.55	1.81E+08	2.12E+07	14.57	523.45
Est. Dev.	2.9	0.44	0.112	221	0.1	48.1	4.2	42.4	8.0	146.4	17.4	35.3	1.1	1.84	8.27	1.55E+08	3.44E+07	12.92	328.12
75 Percentile	28.6	8.08	0.245	1636	0.1	119.4	38.7	278.8	31.3	880.8	138.1	115.8	3.7	3.20	23.15	2.80E+08	4.40E+07	20.90	753.04
25 Percentile	24.3	7.40	0.136	1487	0.0	82.5	35.3	224.4	26.3	756.0	115.6	91.0	2.2	0.03	12.79	8.13E+07	6.98E+06	12.74	364.14
n	4	4	3	4	4	4	4	4	4	4	4	4	4	4	4	4	4	4	4

SEC = Electrical conductivity (μmhos/cm)
CT = Total coliforms
OG = Oil and grease
Eh = redox potential (Volts)
CF = Fecal coliforms
COD = Chemical oxygen demand
DO = dissolved oxygen (mg L^{-1})

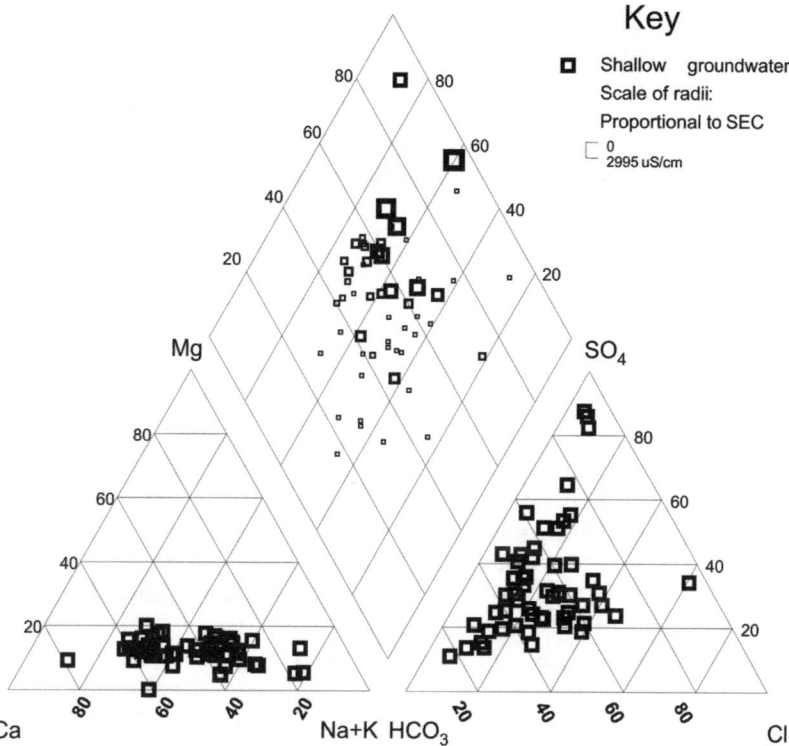

Figure 2. Piper diagram for collected samples in the study area.

2,995–158 μmhos/cm for the first sampling and 3,190–615 μmhos/cm in the second sampling; Tanque Tenorio waste-water showed values between 1,940–1,476 μmhos/cm. Lower salinity values were found at the edges of the SSM in the west, increasing to the east, towards SGS (up to 2,000 mgL^{-1}). Results of chemical analyses show large variations in chemical composition and also indicate zones with relative high salinity of the shallow groundwater. High salinity values in groundwater were found below the crop irrigation zone where raw waste-water is used and adjacent to Villa de Pozos (east part of the industrial park). Zones with high SEC, SO_4, Cl, salinity, correspond well with those previously reported for the shallow aquifer (Carrillo and Armienta, 1989; CNA, 1994) but their concentration and extent have grown substantially as demonstrated in this study.

Most groundwaters are Ca-HCO_3, Na-HCO_3, mixed cation type-HCO_3 or Na-mixed anion type (figure 2). Considering results from table 1, the second sampling shows average concentrations considerably higher than those documented for the first sampling. This probably reflects dilution by infiltrated water derived from precipitation just before the first sampling and/or sampling bias, as the second sampling was directed to the most contaminated zone within the shallow aquifer as identified from first sampling results. Sulphate and Cl concentrations above national drinking water standards were detected in 7% of the samples. Sulphate concentrations reach up to 1,534 mgL^{-1} (second sampling), but about 75% of the samples show values lower than 236 mgL^{-1}. Groundwater with the highest concentration approached (but did not reach) saturation with respect to gypsum. Maximum Cl values are close to 450 mgL^{-1}. Considering that the geologic units of the study area are not an abundant natural source for this element, this high concentration could reflect an anthropogenic source; ≈50% of the sampling area had concentrations below 60 mgL^{-1}, which are considered to be the natural baseline. The concentrations of Cl and SO$_4$ for both first and second sampling as well as available data from precipitation and waste-water are presented in figure 3. Chloride values below and around 10 mgL^{-1} are identified at the edge of SSM, they represent local recharge under semi-arid condition (rainfall composition modified by evapotranspiration). Considering the average Cl value determined in local precipitation to represent the average for modern rainfall (Cl value of 0.61 mgL^{-1}; Cardona and Carrillo-Rivera, 2005) and allowing for Cl in the local recharge zone to be all rainfall-derived, the *minimum recharge rate* in this area would be about 10% of the rainfall (this amounts to ≈40 mm yr^{-1}). Global characterization of atmospheric dust in the SLP area indicates quartz, calcite, fluorite and anhydrite as main mineral components (Campos-Ramos, 2005). The average chemical composition of rainfall presented in Cardona and Carrillo-Rivera (2005) indicates a Ca/SO$_4$ weight ratio (0.51) which is quite similar to the theoretical ratio derived from anhydrite (0.42) dissolution. Average pH of local rainfall is 5.8, which suggest that industrial derived sources for SO$_2$ to the atmosphere are considered to be of minor importance and indicates that the most probable source for the bulk aerosol in local rainfall is dust with some anhydrite content. This source is expected to be from unprotected residues of industrial processes from the industrial park of SLP. Sulphate concentrations in the recharge zone are between 20 and 110 mgL^{-1} and represent local rainfall concentrated during the infiltration process. The Cl-SO$_4$ relationship presented in figure 3 indicates that chemistry of local flow systems

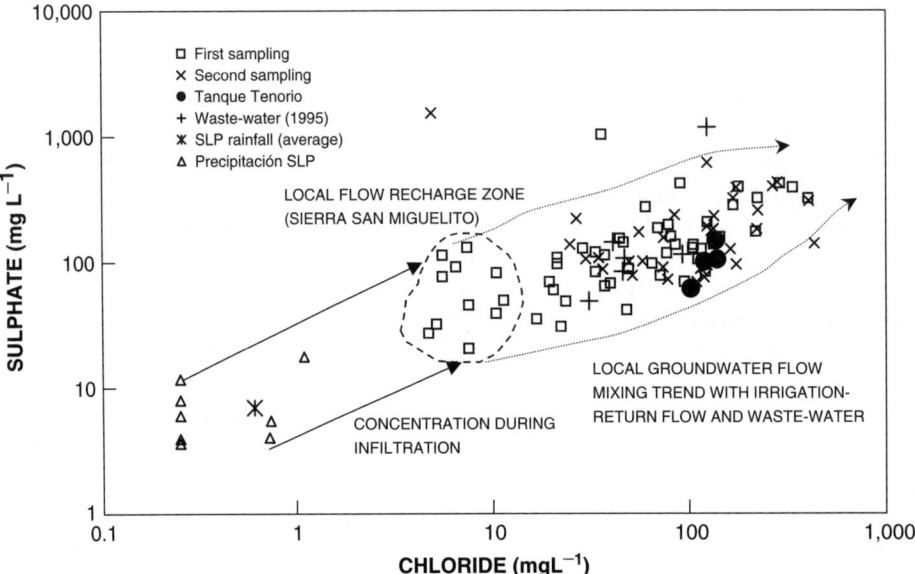

Figure 3. Variation of SO4 with Cl concentration in the Tanque Tenorio waste-water, groundwater (first and second sampling), waste-water (historical data), and precipitation samples collected for this study (average precipitation was taken from Cardona y Carrillo-Rivera, 2005).

(transit zone) within the SLP-SGS zone, is controlled by mixing with waste-water infiltrated as irrigation-return effluents and losses from sewers and open channels delivering waste-water within the city to irrigation areas. Domestic activities introduce some Cl and SO_4 in the water supply used by the inhabitants, resulting in the concentrations detected in waste-water. Tanque Tenorio represents waste-water affected by some evaporation within the open reservoir, producing higher concentration of both Cl and SO_4 than those detected in the original waste-water composition. In some zones, shallow groundwater quality has higher Cl and SO_4 concentrations than those values identified in waste-water. Taking into consideration that there are not ubiquitous and/or abundant natural sources in the basin fill sediments for these elements, a significant concentration of waste-water in irrigation-return effluents during irrigation activities is considered to be the cause.

4.2 Factor analyses

The interrelationships between measured parameters were also investigated with a simple correlation analysis. Traditional correlation coefficients (Pearson's correlation coefficient) were used to measure and test the intensity of the linear relation between two parameters using the covariance of the compared variables, standardized by the standard deviations. Considering the high correlation coefficients in the correlation matrix for major elements and SEC (table 2) it was supposed that R-mode factor analyses could be used to reduce the pattern of correlations between parameters to simpler sets of factors, whose interpretation is believed to be straightforward. The factor analysis method is described elsewhere (Brown, 1998; Davies, 1987; Drever, 1997; Dillon and Goldstein, 1984; Johnson and Wichern, 1992), it consists of several steps including the transformation of data (standardized with mean = 0 and standard deviation = 1), calculation of eigenvalues and eigenvectors, transformation into factors, rotation of factor loading matrix according to a varimax scheme, and producing a new factor loading matrix that is used for the interpretation.

Using factor analysis (R-mode), the major element distribution in the samples of the first sampling is explained in terms of three factors. The selected factors explain more than 82% of the total variance of the data set. The rotated varimax matrix is shown in table 3, it can be seen that the communalities of all the parameters are >0.66, indicating that the 3-factors procedure explains adequately the variance of almost all the parameters.

Table 2. Correlation matrix of chemical parameters and SEC for groundwater (first sampling) in the San Luis Potosi shallow groundwaters. In bold, significant values at the level of significance alpha = 0.050 (two-tailed test).

	Ca	Mg	Na	K	HCO_3	Cl	SO_4	$N-NO_3$	SEC
Ca	1.000								
Mg	**0.752**	1.000							
Na	**0.584**	**0.576**	1.000						
K	**0.468**	**0.362**	**0.831**	1.000					
HCO_3	**0.551**	**0.777**	**0.674**	**0.428**	1.000				
Cl	**0.808**	**0.643**	**0.719**	**0.602**	**0.553**	1.000			
SO_4	**0.792**	**0.559**	**0.525**	**0.463**	0.258	**0.435**	1.000		
$N-NO_3$	**0.614**	**0.712**	**0.578**	0.303	**0.635**	**0.668**	0.310	1.000	
SEC	**0.907**	**0.762**	**0.849**	**0.730**	**0.671**	**0.902**	**0.704**	**0.675**	1.000

Table 3. Loadings for the varimax rotated 3-factors model.

Variable	Factor			Communality
	I	II	III	
Ca	0.483	0.210	0.850	0.999
Mg	0.779	0.143	0.442	0.822
Na	0.470	0.795	0.252	0.916
K	0.140	0.885	0.255	0.867
HCO_3	0.792	0.314	0.101	0.736
Cl	0.553	0.455	0.443	0.709
SO_4	0.130	0.274	0.775	0.693
$N-NO_3$	0.751	0.183	0.245	0.657
SEC	0.550	0.565	0.616	0.999
Total variance %	32.167	24.583	25.477	
Cumulated %	32.167	56.750	82.227	

Factor I accounts for 32.2% of total variance with high loadings in Mg, HCO_3, Cl, $N-NO_3$ and SEC; factor scores above zero have a distribution corresponding with the zone where irrigation practice with raw waste-water is common. This indicates that the SGS zone is affected by the processes represented by this factor, which can be associated with diffuse contamination derived from waste-water usage. Factor II accounts for 24.6% of the total variance with high loadings in Na, K and SEC; this factor can be associated with dissolution of K-feldspar and cation exchange during the infiltration of waste-water, which increases the concentration of Na and K. Factor III accounts for 25.5% of the total variance with high loading in Ca, SO_4 and SEC; distribution of high positive factor scores is similar to that of Factor I, suggesting evapotranspiration of waste-water during the irrigation practices is taking place, producing the concentration of salt content in the irrigation-return flow; the presence of anhydrite in the atmospheric dust in the region can also contribute to these parameters.

4.3 Nutrients

Two inorganic compound indicators were used to show the extent of modern contamination in the local flow systems investigated: N species (NO_3 and NO_2) and P. Waste-water usually exhibits a significant N load including different N-compounds such as NH_4, NH_3, NO_3, NO_2, organic-N. Effluents of a typical septic tank system have a total N content of 25–60 mgL^{-1}; 20–55 mgL^{-1} exists as NH_3 and less than 1 mgL^{-1} as NO_3 (Canter, 1997). Tanque Tenorio waste-water has lower NO_3 concentrations than those detected in groundwater suggesting N load is represented by NH_3 and organic-N species (31.3 and 21.8 mgL^{-1}, respectively (CNA, 1995). Ammonia and NH_4 are easily oxidized within a few hours to a few days and within vertical distances of a few tens of centimetres in the soil (Barrett et al., 1999; Robertson and Blowes, 1995). Nitrification refers to the biological oxidation of NH_4, this is consummated in two steps: reaching first to the NO_2 form, then to the NO_3 form. The transformation reactions are coupled and proceed rapidly to the NO_3 form according to the following reaction.

$$NH_4^+ + 2O_2 \rightarrow NO_3^- + 2H^+ + H_2O$$

In cases where vadose zone residence time is short (hours or less than 1 week) the oxidation processes may remain incomplete producing low NO_3 and DO concentrations

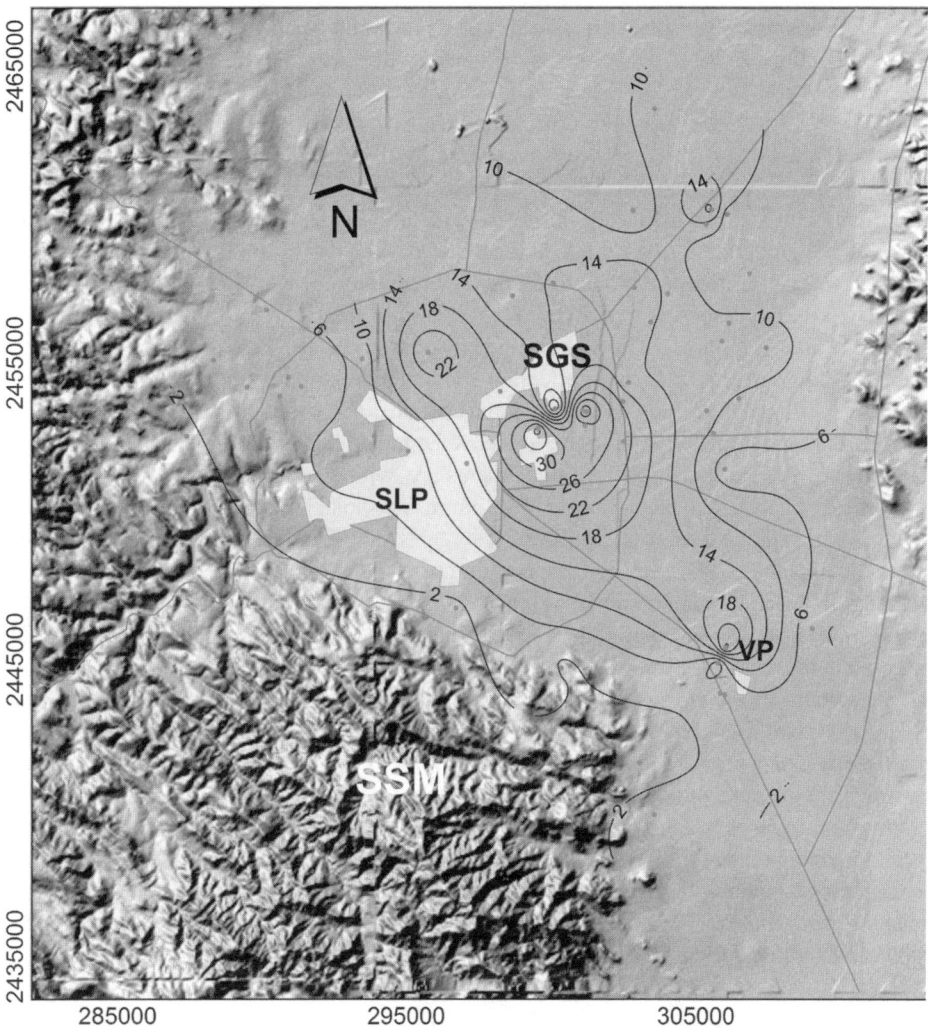

Figure 4. Contour map of iso-concentration lines for N-NO3 (mgL^{-1}) in shallow groundwater.

(Verstraeten et al., 2005); however this should not be the case in the investigated area as DO was present in most of the sampling sites, suggesting nitrification reactions during waste-water infiltration as a major source for NO_3 in shallow groundwater.

Under the oxidizing conditions prevailing in the shallow aquifer, the maximum observed concentrations of both NO_3 (up to 1,920 mgL^{-1} as NO_3) and NO_2 (up to 9.2 mgL^{-1} as NO_2) produced as the intermediate step in the nitrification reaction, are often high and correspond in general with high salinity zones in SGS. In contrast low values of NO_3 and NO_2 are found in the western and south part of the area, close to the local recharge zone of SSM, suggesting a minor impact of diffuse contamination derived from waste-water management (figure 4). Denitrification of NO_3 in groundwater is an advantageous reaction in groundwater, because NO_3 is reduced to N_2, which is not detrimental to drinking water (Freeze y Cherry, 1979). A decline in redox potential of the groundwater can, in some situations, cause denitrification, which would occur at Eh ≈ 0.25 V (pH = 7.0 and temperature 25°C); DO in this environment will be almost nil. Field conditions (Eh above 0.3 V and average DO content of 3.3 mgL^{-1}) for shallow groundwater within the study area suggest that the capacity for such denitrification could be limited and the only way to diminish NO_3 concentrations would be by dilution via hydrodynamic dispersion along groundwater flow. About 55% of the collected samples during the first sampling and almost 70% of samples collected in the second sampling show concentrations exceeding the national guideline for NO_3 in drinking water of 44.3 mgL^{-1} or 10 mgL^{-1} N-NO_3. The toxicity of NO_3 to humans is due to the body's reduction of NO_3 to NO_2 (Canter, 1997), a reaction taking place in the saliva (all ages) and in the gastrointestinal tract of infants (age less than 3 months). The toxicity of NO_2 has been demonstrated by vasodilatory/cardiovascular effects (at high dose levels) and methemoglobinaemia (at low dose levels).

Phosphorus and N are elements in the same chemical group, both can occur in a number of valence states, but for P in most water systems the most fully oxidized (+5) is the common state (Hem, 1989; Fetter, 1999). Both elements are component of sewage because they are almost always present in animal waste. Phosphorus can be liberated to the environment from a number of additional sources (fertilizers, detergents, etc.). Waste-water within the city in the sewer network has a PO_4 average concentration of 27 mgL^{-1} (CNA, 1995), raw waste-water feeding the reservoir shows similar values (22.7 mgL^{-1}); concentrations in Tanque Tenorio waste-water are lower (average 17.4 mgL^{-1}) than those values. The reason for decreasing values within the reservoir is probably due to vegetation fixation of P as a nutrient. The detected concentrations in groundwater are much lower, suggesting effective attenuation processes in the vadose zone. Geographic distribution of PO_4 in the saturated zone as evidenced from the first sampling is presented in figure 5. Values below 0.05 mgL^{-1} were found below the main crop irrigation zone; within the SLP City zone concentrations are generally higher. No relationship between NO_3 and PO_4 concentrations is evident. The behaviour of PO_4 in groundwater as compared with the conservative nature of NO_3 in the system indicates that the main controls for its mobility are likely to be adsorption to soil components (oxidizing conditions probably contribute to PO_4 adsorption on ferric and manganese oxy-hydroxides) and vegetation consumption during application of waste-water to crop irrigation.

4.3 *Microorganisms*

A large number of microorganisms known to be pathogens can be found in groundwater. Their presence in the subsurface is generally associated with the introduction of faecal material present in different sources of contamination; waste-water probably represents the main source as it contains viral, bacterial and protozoan pathogens. The listing of microbial pathogens in groundwater includes more than 100 viral, bacterial pathogens and protozoa (Macler and Merkle, 2000). Although many coliform bacteria are not pathogenic, the presence in groundwater of thermotolerant coliform bacteria is often used as an indicator of the likelihood of the presence of other pathogenic bacteria, and virus. Total coliform bacteria, faecal coliform bacteria, *E. coli* and coliphages can be indicators of waste-water contamination of groundwater (Verstraeten *et al.*, 2005). In this study, in order to get additional indicators of waste water impact, total and faecal coliforms were considered. Total coliform counts are only helpful when used in conjunction with faecal counts, because the former can be derived from natural populations in soils and then be found in rainfall recharge and lack any waste-water significance. Considering the Mexican Standard (NOM-001-ECOL-1996) for waste-water

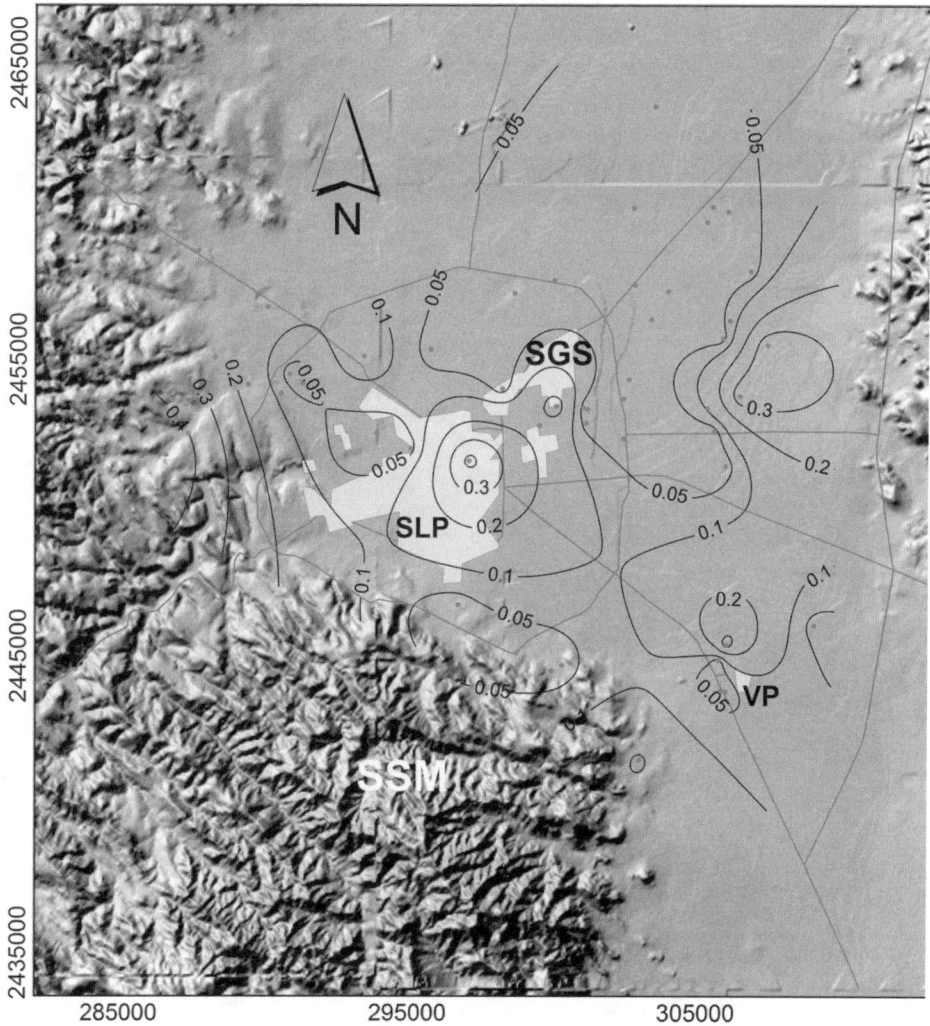

Figure 5. Contour map of iso-concentration lines for PO_4 (mgL^{-1}) in shallow groundwater.

discharge to open reservoirs (total coliforms values lower than 1,000 cfu 100 mL^{-1}) local waste-water is actually not suitable for direct discharge. Tanque Tenorio waste-water has 2.55×10^8 and 4.95×10^2 cfu 100 mL^{-1} for total and faecal coliforms, respectively, while waste-water before entering the reservoir shows higher values $3.55 \times 10^8 - 3.3 \times 10^7$ cfu 100 mL^{-1}. These conditions suggest that Tanque Tenorio is working as a rudimentary treatment plant, oxidizing the waste-water and diminishing especially the faecal coliform content; however, its efficiency is not adequate to completely mitigate the contamination of the environment. The values of total coliforms in groundwater are at least three orders of magnitude lower (between 1.95×10^5 and 1×10^1 cfu 100 mL^{-1}) but the modest log-reduction in counts still suggest a biological impact of waste-water disposal. The identified total and faecal coliforms content difference between waste-water and shallow

Figure 6. Contour map of iso-concentration lines for total coliforms (cfu 100 mL^{-1}) in shallow groundwater.

groundwater suggests that most micro-organisms are removed during infiltration of wastewater in the top few meters of the vadose zone, but deep micro-organisms penetration during the dry season may develop because the activity of native soil bacteria is lower, producing a less antagonistic media in the soil. Total coliforms distribution in groundwater is presented in figure 6, a major impact is evident below the crop irrigation zone adjacent to SGS and around the industrial park (Villa de Pozos). Local biological contamination in shallow hand-dug wells is expected as field observations indicate inadequate conditions in their design, construction and operation, most of them functioning under non-sanitary conditions.

4.4 Organic pollution load

The term oils and greases (OG) applies to a wide variety of organic substances extracted from aqueous solution or suspension by hexane; it includes hydrocarbons, esters, oils, fats, waxes and high-molecular-weight fatty acids. Grease generally originates from meat, seeds and some fruits; it is not easily degraded by microorganisms. Another common type of oils include kerosene, lubricating oil, asphalt, compounds used in different types of industry and generated in road runoff. Tanque Tenorio waste-water has an average OG concentration of 19.1 mgL^{-1}, concentration in waste-water along collecting channels is higher (60–70 mgL^{-1}; CNA, 1995); both of these values are higher than those detected in shallow groundwater (about 1.5 mgL^{-1}); however, in some sites OG concentration was below detection limit. The highest values of OG in groundwater were identified below the main crop irrigation zone adjacent to SGS, as well as in Villa de Pozos close to the industrial park; additional sites within SLP City with OG concentrations represent the influence of leaks from the piped sewer network and infiltration from unlined waste-water collection channels.

The COD content is a measure of the amount of oxidizable organic matter present in the water, it is used here to represent the pollution load derived from waste-water; it has no particular direct geochemical significance but was useful in this study as a measure of the gross organic contaminant load derived from waste-water during infiltration to shallow groundwater. Tanque Tenorio waste-water is highly prone to oxidation; COD values have an average of 593 mgL^{-1} suggesting a pollution load applied to soil during irrigation could produce a negative impact. Values identified in shallow groundwater are much lower, ranging from 9.5 to 22.8 mgL^{-1} for the first and the second sampling, respectively. Oxidation of organic pollution load in the vadose zone, as well as dilution by hydrodynamic dispersion in the saturated zone, are likely to be the main processes responsible for the observed decrease in value of COD during the infiltration of waste-water. Relatively high values of COD were detected at sites on one part of the recharge zone. Together with low NO$_3$ and salinity as well as a lack of irrigation with waste-water in this region, it is inferred that they represent a local (point) contamination problem as poor sanitary conditions prevailing around the hand-dug well, together with an inappropriate design and construction contribute to the organic contamination of shallow groundwater. Samples taken within the SLP City showed COD values higher than 5 mgL^{-1}, suggesting leakage of the sewer system that has travelled towards the shallow groundwater.

5 CONCLUSIONS

Based on results of analyses of several indicators, including major anions (Cl, SO$_4$), nutrients (NO$_3$, NO$_2$, PO$_4$), faecal coliforms, OG and COD, it becomes evident that the shallow groundwater in the SLP City area has a range of quality problems. The most serious includes salinity and high concentration of some nutrient and micro-organisms; heavy metals were not investigated but it is reasonable to anticipate that they may be present in the system. The evidence of waste-water contamination was especially strong because multiple indicators of sewage contamination were detected in the same sample; for example high Cl, SO$_4$, NO$_3$ and NO$_2$ concentrations together with presence of faecal coliforms. The shallow aquifer was identified to be quite vulnerable to waste-water contamination; the present irrigation practices and waste-water management increase susceptibility to the infiltration of irrigation return-flow. The combined evidence provides indication that there is a significant human impact on groundwater quality of the shallow aquifer, producing a noticeable deterioration in the overall quality of the groundwater resource and seriously degrading it for drinking water purposes without treatment. This suggests that historical and present-day waste-water management has substantially changed the natural baseline conditions in the central part of the study area; however, the low salinity and NO_3 zone adjacent to SSM probably gives an indication of the groundwater quality that is close to the natural baseline.

While the vertical hydraulic gradient shows high values, vertical hydraulic conductivity for the compact sandlayer is expected to be lower than the measured horizontal hydraulic conductivity (10^{-9} ms^{-1}), thus limiting the amount of infiltration water from shallow contaminated zones to the deep aquifer. This hydrogeological control gives natural protection, to some extent, to underlying water resources in intermediate and regional flow systems that are tapped for drinking water purposes. However, there are many poorly constructed hand-dug wells and boreholes (inactive and active) in the area that connect the local flow system with the intermediate and regional systems, allowing shallow contaminated water to pollute deep sources of potable water supply. Such condition constitutes a major threat to the sustainability of the deep drinking water sources in the area. The successful definition of the groundwater flow system at both local and regional scale (Carrillo-Rivera *et al.*, 1996) has allowed the physicochemical character of the different flows to be interpreted. This understanding has provided a valuable tool to investigate how the upper local (waste-water contaminated) flows have evolved in the shallow aquifer, and how they would, in time, influence the quality of the intermediate and regional flows that are currently used for water supply in the study area.

ACKNOWLEDGMENTS

The authors thank the collaboration of the Laboratory of Chemistry from the Comisión Nacional del Agua, Gerencia Estatal San Luis Potosi, specially to Aracely García and Juana Bertha Leyva. We also recognize the collaboration of the Earth Sciences Water and Soil Chemistry Laboratory staff of Facultad de Ingeniería-UASLP in performing the water chemical analyses. Jorge Aceves from the GIS-Lab (FI-UASLP), helped with the production of the digital elevation model of the study area. Partial financial suppport from Secretaria del Medio Ambiente y Recursos Naturales (Grant SEMARNAT-2002-C01-0719) and Universidad Autónoma de San Luis Potosí, Secretaria de Investigación y Posgrado (Grant C03-FRC-07-2.19) is also recognized.

REFERENCES

Barrett MH, Hiscock KM, Pedley S, Lerner DL, Tellam JH and French MJ (1999) Marker species for identifying urban groundwater recharge sources: A review and case study in Nottingham, UK. Water Research 33: 3083–3097

Brown CE (1998) Applied multivariate statistics in geohydrology and related sciences, Springer-Verlag

Campos-Ramos AA (2005) Caracterización de las partículas contenidas en el polvo atmosférico en el entorno de la zona industrial de San Luis Potosí, Master Thesis, Ingeniería de Minerales, Facultad de Ingeniería, UASLP, México

Canter LW (1997) Nitrates in Groundwater, CRC Press

Cardona A and Carrillo-Rivera JJ (2005) Hidrogeoquímica de sistemas de flujo intermedio que circulan por sedimentos continentales derivados de rocas volcánicas félsicas. Ingeniería Hidráulica en México, In press

Carrillo-Rivera JJ (1992) The hydrogeology of the San Luis Potosi Area, México. Ph.D. thesis, University of London, U.K.

Carrillo-Rivera JJ and Armienta MA (1989) Diferenciación de la contaminación inorgánica de las aguas subterráneas del valle de la ciudad de San Luis Potosí, México. Geofísica Internacional 28(4): 763–783

Carrillo-Rivera JJ, Cardona A and Moss D (1996) Importance of the vertical component of groundwater flow: a hydrochemical approach in the valley of San Luis Potosí, Mexico. J. Hydrol 185: 23–44

Carrillo-Rivera JJ, Cardona A and Edmunds W (2002) Use of abstraction regime and knowledge of hydrogeological conditions to control high-fluoride concentration in groundwater: San Luis Potosí basin, México. Journal of Hydrology 261(1–4): 24–47

Chilton PJ, Morris BL and Foster SSD (1996) Impacto del reuso de las aguas residuales sobre el agua subterránea en el Valle del Mezquital, Estado de Hidalgo, México. in VIII Curso Internacional Los Recursos Hídricos subterráneos y la disposición de aguas residuales urbanas, interacciones positivas y negativas, 36–46, Querétaro, México

CNA (Comisión Nacional del Agua) (1994) Proyecto piloto de contaminación del acuífero de la zona industrial de San Luis Potosí, S.L.P. Internal Technical Report, Gerencia Estatal San Luis Potosí, Subgerencia Técnica

CNA (Comisión Nacional del Agua) (1995) Problemática de contaminación en la zona industrial. Internal Technical Report, Gerencia Estatal San Luis Potosí, Subgerencia Técnica

Davies JC (1987) Statistics and analysis in Geology, 2nd edn, John Wiley and Sons, New York

Dillon WR and Goldstein M (1984) Multivariate analysis. Methods and applications. John Wiley & Sons, New York

Domínguez-Mariani E, Carrillo-Chávez A, Ortega A and Orozco-Esquivel MT (2004) Waste-water reuse in Valsequillo Agricultural Area, Mexico: environmental impact on groundwater. Water, Air and Soil Pollution 155: 251–267

Drever JI (1997) The geochemistry of natural waters, 3rd edn, Prentice Hall Inc.

Fetter CW (1999) Applied hydrogeology. 2nd edn. Merril Publishing Company.

Freeze RA and Cherry JA (1979) Groundwater. Prentice Hall Inc.

Geoingeniería Internacional (1996) Estudio Hidrogeológico de la contaminación del agua subterránea en la zona industrial de San Luis Potosí, S.L.P., estudio realizado para la Comisión Nacional del Agua, Gerencia de Aguas Subterráneas

Hem JD (1989) Study and interpretation of the chemical characteristics of natural water, Third edition, U. S. Geological Survey, Water Supply Paper 2254

Hvorslev MJ (1951) Time lag and soil permeability in groundwater observations. U.S. Army Corps Engs. Waterways Exp. Sta. Bull. 36, Vicksburg, Miss.

Johnson RA and DW Wichern (1992) Applied multivariate statistical analysis. Prentice-Hall, Englewood Cliffs

Labarthe-Hernández GM, Tristán-González M and Aranda-Gómez JJ (1982) Revisión estratigráfica del Cenozoico de la parte central del Estado de San Luis Potosí. UASLP, Instituto de Geología, Folleto Técnico 85

Macler BA and Merkle JC (2000) Current knowledge on groundwater microbial pathogens and their control. Hydrogeology Journal 8: 29–40

Molina-Maldonado A, Cardona A and Marín SL (2001) Modelo hidrogeoquímico conceptual de la Comarca Lagunera. Memorias del XI Congreso Nacional de Geoquímica, Ensenada Baja California

Robertson WD and Blowes DW (1995) Major ion and trace metal geochemistry of an acidic septic-system plume in silt. Groundwater 33: 275–283

Silva-Lugo H (2005) Evaluación del impacto producido por actividades antropogénicas en acuíferos costeros de zonas áridas: Costa de Hermosillo, Sonora. BS Thesis, Ingeniero Geólogo, Facultad de Ingeniería, UASLP, México

Steinich B, Escolero O and Marin LE (1998) Salt-water intrusion and nitrate contamination in the Valley of Hermosillo and El Sahuaral coastal aquifers, Sonora, Mexico. Hydrogeology Journal, 6(4): 518–526

Stretta EJP and Del Arenal R (1960) Estudio para el abastecimiento de agua potable para la Ciudad de San Luis Potosí. Applied Sciences Institute, Hydrology Section, UNESCO and Instituto de Geofísica, UNAM. Internal Report, January

Tóth J (1995) Hydraulic continuity in large sedimentary basins, Hydrogeology Journal 3(4): 4–16

Verstraeten IM. Fetterman GS, Meyer MT, Bullen T and Sebree SK (2005) Use of tracers and isotopes to evaluate vulnerability of water in domestic wells to septic waste. Groundwater Monitoring and Remediation 25: 107–117

CHAPTER 4

Integrating physical hydrogeology, hydrochemistry, and environmental isotopes to constrain regional groundwater flow: southern Riverine Province, Murray Basin, Australia

Ian Cartwright[1], Tamie R. Weaver[2] and Sarah O. Tweed[1]
[1]*Hydrogeology and Environment Research Group, School of Geosciences, Monash University, Clayton, Vic., Australia*
[2]*Hydrogeology and Environment Research Group, School of Earth Sciences, University of Melbourne, Parkville, Vic., Australia*

ABSTRACT: The hydrochemistry of groundwater in the southern Riverine Province of the Murray Basin is controlled largely by evapotranspiration with minor halite, silicate, carbonate, and gypsum dissolution. Groundwater salinity is highly variable (total dissolved solids, TDS, contents <100 to ~50,000 mgL^{-1}) and controlled by recharge rates. In the palaeovalleys of past streams ("deep leads") more rapid recharge through coarser-grained sediments of the unconfined Shepparton Formation produces lower salinity groundwater in both shallow and deeper aquifers. Away from the deep leads, recharge rates are slower through the more clay-rich Shepparton Formation sediments. The distribution of salinity, trends in major ion and stable isotope ratios, and the distribution of percent modern carbon contents imply that groundwater flow in the southern Riverine Province is locally complex. Vertical flow occurs within the Shepparton Formation. In the deep leads, there is significant lateral flow in the deeper Calivil-Renmark Formation, while in the intermediate areas, there is much greater relative vertical leakage through the Shepparton Formation into the Calivil-Renmark Formation. Despite near-surface processes largely controlling groundwater chemistry, and climate in southeast Australia varying over the last 25–30 ka, there is little difference in chemistry between the older deeper groundwater of the Calivil-Renmark Formation and younger shallower groundwater from the Shepparton Formation. Recent land clearing has increased recharge causing the water table to rise. Long-term this may result in the groundwater in the Murray Basin becoming less saline, although in the short-term the rise of pre-existing saline water towards the surface represents a major environmental problem.

Keywords: groundwater, geochemistry, Murray Basin, environmental isotopes, 14-C.

1 INTRODUCTION

In a semi-arid region, such as southeast Australia, groundwater is the most valuable long-term water resource. A significant part of rural southeast Australia depends on groundwater from the Murray Basin for agricultural, industrial, and, increasingly, domestic water supply. This demand will increase as population grows, and ongoing development of this region relies on the long-term sustainable use of groundwater. Land clearing over the last 200 years following European settlement has resulted in increased recharge leading to land salinisation and waterlogging, degradation of wetlands and rivers, and soil erosion (e.g. Allison et al., 1990; Ghassemi et al., 1995). This has already degraded groundwater and surface water in this region, threatening water supplies, ecosystems, agricultural productivity, and the cultural value of land. Sustainable use of groundwater requires a thorough knowledge of basin hydrogeology, including groundwater flow patterns, origins of solutes, and groundwater-surface water interactions both under present day and pre-land clearing conditions. This paper uses hydrochemistry to describe long-term regional groundwater flow paths and major hydrological processes in the southern Riverine Province of the Murray Basin, Australia.

1.1 Hydrogeology of the southern Riverine Province

The Murray Basin (figure 1a) occupies $\sim 3,00,000$ km^2 of southeast Australia and contains a series of late Palaeocene to Recent sediments that overlie Proterozoic to Mesozoic basement. The geology and hydrogeology of the Murray Basin are discussed by, amongst others, Tickell (1978, 1991), Tickell and Humphries (1986), Arad and Evans (1987), Calf et al. (1986); Lawrence (1988), Stephenson and Brown (1989), Macumber (1991); Love et al. (1993), Herczeg et al. (1993, 2001), Dogramaci and Herczeg (2002), Cartwright and Weaver (2005); and Cartwright et al. (2004, 2006). At its deepest, the basin is up to 600 m thick; however, the majority of the basin is <400 m thick. The basin is divided into three sub-basins or provinces (Riverine, Scotia, and Mallee-Limestone: figure 1a). This division is on the basis of groundwater flow patterns that are controlled by the presence of hydrological barriers such as basement ridges (figure 1b). Except for a small region in the west that discharges to the Southern Ocean, the basin is closed and groundwater discharges to salt lakes near the basin centre. The River Murray and its tributaries are the only major surface water features draining the basin (Herczeg et al., 1993).

The southern Riverine Province underlies the Riverine Plain of Victoria and New South Wales (figs 1 and 2). It is separated from the Scotia Province by the Neckerboo Ridge, and from the Mallee-Limestone Province by a change in groundwater flow direction that coincides with the eastern edge of the Murray Group limestone aquifer and its over- and underlying low-permeability units (the Winnambool Formation, Geera Clay, and Bookpurnong Beds: Lawrence, 1988). The Cainozoic sediments of the southern Riverine Province are dominantly terrestrial with a transition to marginal marine units in the west of the province. The region under discussion in this paper is east of the lower permeability units where the Cainozoic units are terrestrial and where there are no major aquitards. The Murray Basin is unlike many other groundwater basins (e.g., the Milk River aquifer: Hendry et al., 1991; the Great Artesian Basin: Radke et al., 2000; or the East Midlands Triassic: Edmunds et al., 1982) in that many of the basin sediments remain largely unconfined and there are few aquitards (figure 1b). These differences allow recharge to occur across broad areas and make the deeper aquifers vulnerable to contamination from leakage of surface water and shallow groundwater (Arad and Evans, 1987).

There are three main stratigraphic units (figure 1b). The lowermost Renmark Group consists of Palaeocene to late Miocene fluvial clays, silts, sands, and gravels. Overlying the Renmark Formation are the Pliocene sands of the Calivil Formation. In most of the southern Riverine Province, the Calivil Formation is in hydraulic continuity with the underlying Renmark Formation and these formations may be considered as a single,

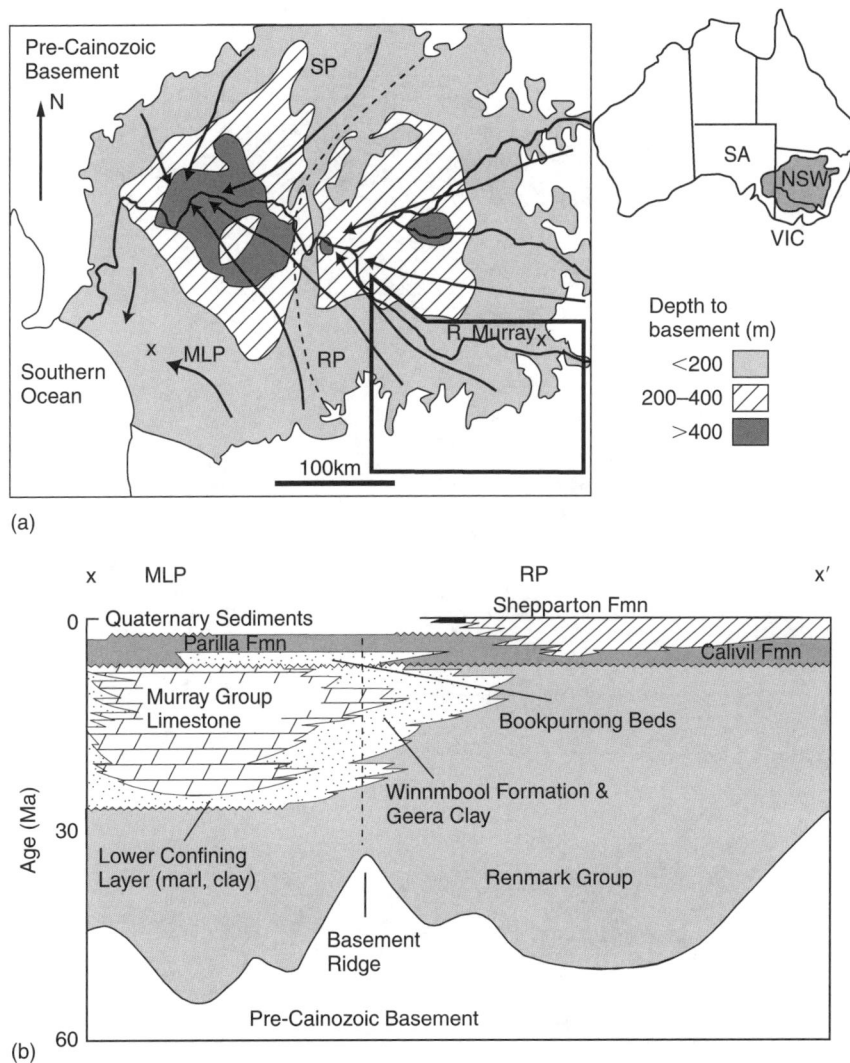

Figure 1. **1a.** Map of the Murray Basin (after Evans and Kellett 1989) showing depth to basement and groundwater flow paths (arrows). MLP = Mallee-Limestone Province, RP = Riverine Province, SP = Scotia Province. Boundary between the Riverine and Scotia/Mallee-Limestone Provinces shown by dashed line; boundary between the Scotia and Mallee Limestone province is the River Murray. Inset shows location of the Basin in New South Wales (NSW), South Australia (SA), and Victoria (VIC). Box shows location of figure 2. **1b.** Stratigraphic cross-section between x and x' (figure 1a) showing major units in the Murray Basin (after Evans and Kellett, 1989).

semi-confined, aquifer (Lawrence, 1988; Macumber, 1991). The Calivil-Renmark Formation does not crop out and is everywhere overlain by the Shepparton Formation. Except at the western edge of the Riverine Province, there are no aquitards separating the Shepparton and Calivil-Renmark Formations (figure 1b). The Calivil-Renmark Formation was deposited by, and is thickest in, ancestral drainage channels ("deep leads") of present day rivers (e.g., the Murray, Campaspe, Lodden, Avoca, and Goulburn Rivers), which were

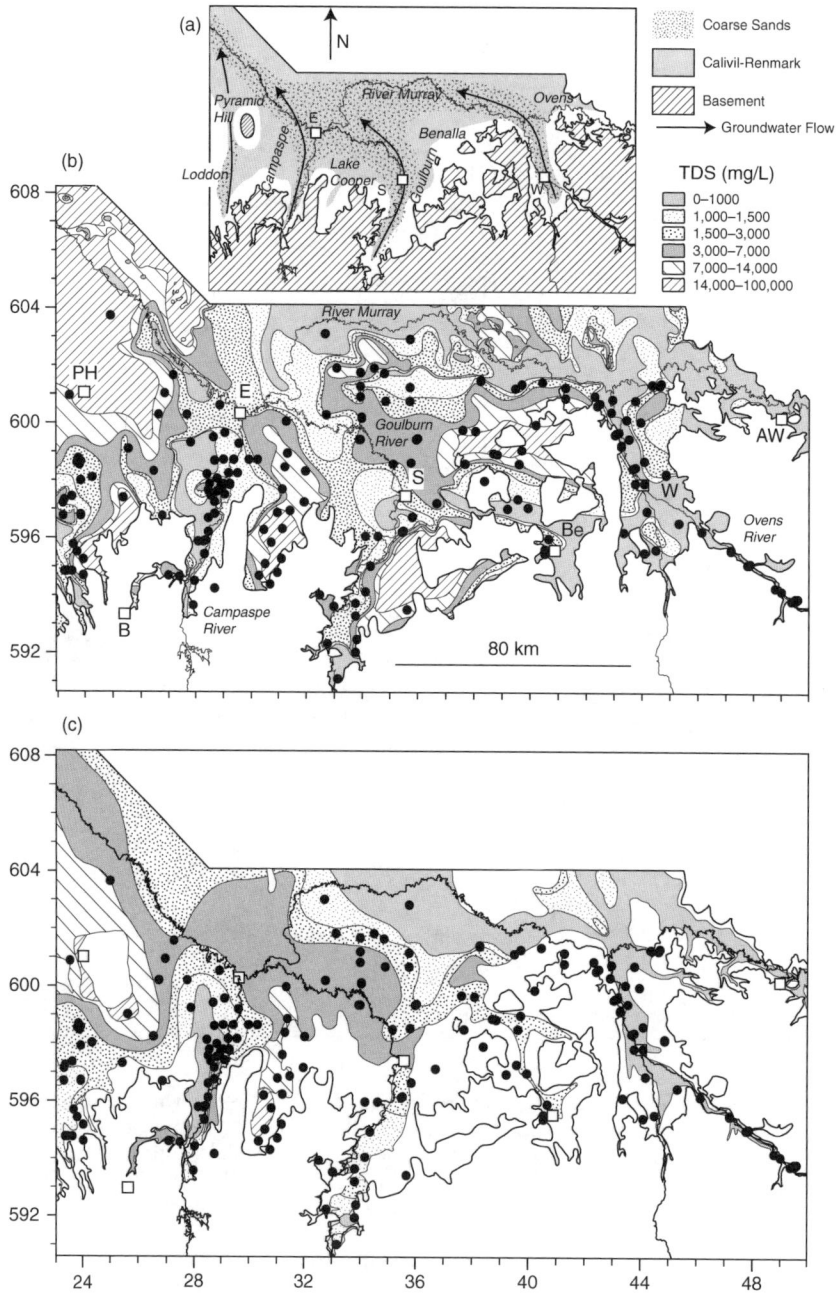

Figure 2. **2a.** General distribution of the deep lead systems in the Calivil-Renmark Formation, groundwater flow paths, and the main river systems in the southern Riverine Province (after Macumber, 1991). **2b,c.** Distribution of TDS contents in groundwater from the Shepparton (**2b**) and Calivil-Renmark (**2c**) in the southern Riverine Province. Data from Tickell and Humphries (1985), Dimos et al. (1994), Hennessy et al. (1994). These distributions represent broad averages and many local variations exist. AW = Albury-Wodonga, B = Bendigo, Be = Benalla, E = Echuca, PH = Pyramid Hill, S = Shepparton, W = Wangaratta. Closed circles show locations of samples reported in this study. Numbers at the map margins are from the Australian Map grid.

established after the Middle Miocene marine regression (Macumber, 1991). Groundwater in the deep leads flows northwards and feeds into the Murray deep lead where groundwater flow is to the west (figure 2). Horizontal hydraulic conductivities of the Calivil-Renmark sediments within the deep leads based on pumping or slug tests are 40 to 200 m/day (e.g., Tickell and Humphries, 1986); slightly lower estimates of 7–60 m/day were made by Calf et al. (1986) and Cartwright and Weaver (2005) from ^{14}C ages. Hydraulic conductivities in the areas between the deep leads are likely to be lower (Tickell and Humphries, 1986).

The uppermost aquifer in the southern Riverine Province is the Shepparton Formation, which comprises a series of fluvio-lacustrine sediments. These sediments include clays, sands, and silts that are laterally discontinuous resulting in a highly heterogeneous aquifer system. Tickell and Humphries (1986) estimated that horizontal hydraulic conductivities are 30 m/day for the coarser units and substantially less in the fine-grained units. Vertical hydraulic conductivities are 10^{-5} to 10^{-1} m/day (Tickell and Humphries, 1986; Cartwright and Weaver, 2005). Recharge of groundwater into the Shepparton Formation occurs across the southern Riverine Province and the heterogeneity may inhibit lateral flow, promoting downward leakage into the underlying Calivil-Renmark Formation.

The southern Riverine Province may be divided into several subcatchments (figure 2). Of these the Ovens, Goulburn, Loddon, and Campaspe subcatchments are typical deep lead systems, while the Benalla, Lake Cooper, and Pyramid Hill regions represent intermediate areas that generally contain more saline groundwater. Annual rainfall in the area depicted in figure 2 varies from ~1,000 mm southeast of Wangaratta to <400 mm in the northwest of the region; most of the region has 400–600 mm annual rainfall (Bureau of Meteorology, 2005). Rainfall occurs dominantly in the austral winter months (July–September) and for much of the year potential evapotranspiration rates exceed rainfall (Bureau of Meteorology, 2005).

Most research into the hydrogeology of the southern Riverine Province has focused on lowsalinity groundwater within the deep leads that represents a potential resource for domestic or agricultural supply. However, as elsewhere in the Murray Basin, large parts of the southern Riverine Province contain groundwater that has total dissolved solids (TDS) contents of >5,000 mgL^{-1} (Evans, 1988; Dimos et al., 1994; Hennessy et al., 1994: figure 2). That the saline groundwater exists across the basin from its margins to the discharge areas implies that it is not simply the result of progressive mineral dissolution during groundwater flow, nor is there a simple correlation with rainfall.

In this paper, we use ^{14}C, major ion, and stable isotope data to document hydrological processes in the southern Riverine Province. Together with groundwater elevations, these data are used to document regional groundwater flow, groundwater recharge, and mixing that have occurred over long time frames. This study illustrates the value of geochemical data in constraining regional flow systems in large basins.

2 DATA SOURCES AND ANALYTICAL TECHNIQUES

Aspects of the hydrogeochemistry of the Loddon, Campaspe, and Goulburn deep leads (figure 2) was discussed by Macumber (1991), Arad and Evans (1987), and Cartwright and Weaver (2005), while Cartwright et al. (2006) discussed Cl/Br ratios and ^{14}C concentrations for groundwater across the southern Riverine Province. Calf et al. (1986) also presented ^{14}C data for a number of bores in the Goulburn and Campaspe deep leads. The new data reported in

Table 1. Geochemistry of groundwater from the southern riverine province of the murray basin.

Bore	East[1]	North[1]	Screen[2] m bns	Aquifer[3]	pH	TDS mgL^{-1}	DO$_2$ mgL^{-1}	CO$_2$ mgL^{-1}	HCO$_3$ mgL^{-1}	Cl mgL^{-1}	Br mgL^{-1}	NO$_3$ mgL^{-1}	SO$_4$ mgL^{-1}	K mgL^{-1}	Na mgL^{-1}	Ca mgL^{-1}	Mg mgL^{-1}	Si mgL^{-1}	SiO$_2$(a)[4] SI	δ^{18}O ‰	δ^{2}H ‰	^{14}C pmc
Benalla																						
88034	368100	5969500	76–94	B	6.92	10900	3	10	bd	7600	14.8	0.2	11.6	16.4	2620	272	326	1.0	−2.00	−5.9	−37	
90121	398800	5988500	85–90	B	7.15	7510	3	80	312	4260	9.8	0.8	714	7.6	1590	287	243	6.2	−1.21	−6.1	−37	
113693	414287	6007428	25–29	B	7.84	6560	6	204	512	3300	6.2	1.6	687	11.0	1310	164	338	22.3	−0.66	−5.9	−35	97
113694	414287	6007428	15–17	B	6.95	5360	7	320	251	2810	4.9	7.6	559	7.6	856	152	358	32.5	−0.5	−5.5	−37	97
138329	403722	5997992	17–20	B	6.1	24100	6	172	202	13900	28.8	3.6	3070	26.5	5150	792	732	50.9	−0.25	−4.1	−29	96
138502	385375	5977627	15–18	B	7.19	2180	2	260	631	609	2.2	11.1	116	0.3	202	200	101	46.5	−0.35	−4.7	−33	
138503	385375	5977627	7–10	B	7.12	2510	2	320	721	703	2.4	0.6	117	0.9	279	223	100	44.7	−0.37	−4.9	−32	
53674	398700	6012300	90–93	C-R	7.48	875	5	74	297	242	0.6	0.1	57	1.9	168	8.1	14.6	9.6	−1.04	−6.5	−38	42
69545	383951	6013760	109–111	C-R	7.63	1020	4	132	200	357	0.8	0.2	69	2.9	216	14.1	22.1	9.3	−1.06	−6.2	−39	27
98130	388364	5987309	74–87	C-R	6.75	3790	1	252	463	1970	4.2	24	8.1	19.5	890	48	87.1	7.4	−1.15	−5.7	−37	72
98131	390070	5987043	83–101	C-R	8.03	3270	3	6	32	2210	5.8	1.1	3.1	10.7	946	18.1	27.6	0.4	−2.42	−6.0	−38	
109356	382200	5995450	82–102	C-R	6.81	3620	4	6	16	2400	5.3	0.8	8.1	9.5	1150	12.5	12.3	0.4	−2.41	−5.6	−36	
109357	377700	5995700	102–104	C-R	6.89	4250	4	100	151	2390	5.1	0.7	433	7.1	951	130	165	21.3	−0.69	−5.5	−36	6
136819	397599	5983734	66–72	C-R	6.92	649	4	24	97	298	0.6	0.8	22.9	3.4	150	15.9	18.8	16.4	−0.81	−3.7	−23	52
48935	408300	5956700	15–20	S	7.74	3140	4	294	132	1680	3.6	0.3	245	9.5	2160	28.4	87	23.6	−0.64	−5.3	−27	
53676	398700	6012300	9–15	S	6.83	7560	3	84	421	3379	7.0	19.2	1530	6.4	1780	109	195	27.5	−0.57	−5.5	−33	
53678	398700	6012400	15–19	S	6.54	9780	8	292	458	4890	9.2	30.2	1550	12.7	1990	134	376	34.5	−0.46	−5.5	−33	99
65845	401100	5967700	24–28	S	6.14	758	4	172	39	336	0.7	2.6	5.1	3.1	129	10.2	25.5	33.9	−0.49	−6.4	−38	99
65846	401100	5967700	8–10	S	5.88	1090	6	60	12	670	1.8	1.9	4.1	3.0	236	15.7	38.2	47.8	−0.34	−6.0	−36	
69547	383950	6013750	74–76	S	7.82	286	8	16	119	35	0.1	0.1	22	0.8	68	1.7	3.3	18.8	−0.75	−6.8	−40	54
69548	383925	6013700	4–22	S	7.57	825	6	116	329	131	0.4	0.4	39	11.9	126	18.7	36.5	14.6	−0.86	−4.0	−26	
98132	388364	5987309	16–19	S	7.41	6800	4	408	792	2970	5.8	41.4	654	8.2	1550	138	205	27.2	−0.58	−5.1	−33	101
105821	378500	5983730	50–77	S	8.01	1760	4	100	395	628	1.9	0.9	165	5.6	370	33.3	53.2	7.1	−1.18	−5.0	−33	
108201	406150	6013150	42–45	S	6.74	3890	4	196	248	1980	3.9	1.6	106	7.7	1130	72.5	123	20.7	−0.7	−6.6	−39	
108202	406150	6013150	14–20	S	8.39	3660	3	514	1030	1170	2.8	60.9	274	92	426	24.6	57.5	11.7	−0.97	−2.1	−9	
108203	406150	6013150	38–40	S	7.14	2660	bd	38	343	1370	2.8	0.2	241	7.6	502	40.6	91	19.3	−0.73	−6.7	−38	
111055	406970	5952900	23–24	S	6.26	605	3	110	29	265	0.6	3.1	10.0	5.2	122	10.4	22.3	25.8	−0.61	−5.9	−36	
113079	406880	5951430	12–13	S	6.54	783	5	14	7	469	1.0	16	11.8	9.2	223	6.7	16	8.2	−1.11	−7.7	−51	
113080	406880	5951430	22–23	S	6.16	159	5	102	15	19	0.1	4.2	2.0	2.3	9.4	2.5	1.8	1.3	−1.91	−9.2	−61	
113691	414245	6011176	27–39	S	7.53	648	8	122	110	175	0.4	2.5	58	2.0	137	5.6	10.7	25.4	−0.62	−7.0	−40	
113692	414245	6011176	4–7	S	8.25	894	5	94	365	147	0.3	4.6	34	35.9	112	27.6	51	19.5	−0.75	−5.6	−34	
114136	396369	6010745	18–24	S	6.67	12300	4	380	604	6610	13.8	12	1370	11.3	2460	226	563	36.8	−0.43	−5.5	−36	

114137	396369	6010745	4–10	S	6.4	24700	2	442	453	12500	26.4	6.4	4450	18.3	5550	313	964	21.3	−0.63	−4.6	−30	
136820	397597	5983736	22–23	S	6.57	2750	5	200	270	1180	2.6	7.7	340	13.2	522	82	94	36.2	−0.46	−4.8	−30	99
138325	393599	5967536	17–20	S	6.69	1700	4	240	370	596	1.2	6.4	18	7.0	387	20.0	29.1	23.8	−0.65	−4.8	−29	
138326	397420	5971091	17–20	S	6.05	6230	4	140	202	3830	7.2	33.4	237	9.4	1240	183	297	41.9	−0.39	−4.7	−32	

Campaspe

65875	280000	5941500	53–64	B	7.03	795	3	82	95.16	368	1.1	1.2	2.7	2.2	187	22.2	30.9	1.8	−1.75	−5.7	−35	
82999	279700	5932600	33–39	B	7.25	2270	4	138	359.9	813	2.1	3.3	282	10.4	497	10.9	145	6.4	−1.21	−5.6	−34	
109648	287390	5938550	40–43	B	6.24	11900	1	166	64.66	7290	20.4	9.6	273	26.4	2980	245	836	1.0	−2.02	−4.9	−31	
47247	294199	5985001	58–73	C-R	10.54	890	2	bd	7	512	1.7	1.5	1.7	9.39	338	4.6	13.5	0.6	−2.23	−5.1	−32	50.6
47253	287100	5984800	123–129	C-R	10.07	868	4	bd	24	533	1.8	1.0	29.8	6.3	229	20.0	22.9	0.1	−3.12	−5.1	−28	
60131	284900	5964200	91–123	C-R	7.21	486	1	124	180.56	71	0.3	0.0	0.87	2.37	81	8.7	14.6	2.9	−1.55	−5.1	−29	
60136	285850	5972050	124–140	C-R	6.99	554	2	56	97.6	219	0.7	0.4	0.32	3.99	138	8.3	28.6	1.2	−1.96	−5.4	−35	
60138	284700	5979900	96–120	C-R	7.53	659	2	16	167.14	226	0.8	1.5	37.6	24.7	155	8.9	17.7	4.5	−1.36	−4.8	−32	
62589	284299	5956000	78–102	C-R	6.65	520	2	212	151	49	0.2	3.8	9.9	1.91	52	10	14.3	15.4	−0.83	−5.3	−33	
65873	270100	5943300	35–39	C-R	6.44	13600	1	214	247.66	7070	11.0	15.5	1720	43.2	3670	73	543	7.7	−1.13	−5.3	−35	
79327	296050	5990900	75–78	C-R	7.21	848	bd	38	112.24	441	1.6	1.3	12.0	5.84	202	11.6	20.8	1.4	−1.88	−5.2	−31	
89576	288440	5978350	84–87	C-R	6.70	1330	6	166	151	336	1.2	1.2	364	6.13	212	27.6	51	16.1	−0.81	−5.1	−35	
102827	289200	6004650	104–108	C-R	6.71	2470	1	122	75.64	1280	4.2	2.0	14.1	13.7	884	4.6	72	0.5	−2.33	−5.6	−26	32.8
102828	289200	6004650	160–167	C-R	6.66	2540	1	120	135	1160	5.1	0.6	145	5	835	72	56	8.2	−1.10	−4.8	−35	
47249	294199	5985001	12–14	S	10.48	2400	2	bd	207	1090	3.7	2.3	188	11.7	870	8.8	20.3	1.7	−1.78	−4.9	−27	
47254	287100	5984800	25–26	S	8.59	8960	3	78	311	8078	13.2	2.6	1320	3.56	2532	187	412	16.2	−0.81	−4.9	−30	
60129	284850	5974200	37–40	S	7.24	474	2	62	106.14	100	0.4	2.9	50	3.46	103	11.2	15.7	18.7	−0.74	−5.1	−33	
60182	284900	5964200	5–17	S	6.73	766	5	85	76.86	319	1.3	4.4	47.0	3.8	154	34.9	28	11.9	−0.94	−5.6	−34	
60184	284700	5979900	4–10	S	6.22	4170	bd	280	75.64	1870	6.9	2.1	588	6.75	962	175	178	35.3	−0.47	−4.1	−28	
62600	284299	5956000	5–17	S	6.72	9270	2	230	276.94	4180	11.6	10	1550	5.19	2410	237	346	18.9	−0.74	−5.5	−32	
65874	270100	5943300	7–9	S	7.21	3230	5	110	322.08	1630	3.1	0.3	185	9.48	751	36.0	168	17.7	−0.77	−5.8	−35	
65876	280000	5941500	28–34	S	6.64	1350	2	58	63.44	586	2.0	22.0	166	2.41	321	40.9	74.6	14.9	−0.84	−5.6	−35	
79328	296050	5990900	28–30	S	7.15	1080	1	214	390.4	207	0.7	0.5	0.9	12	209	10.9	14.7	21.3	−0.69	−5.7	−33	
89594	288440	5978350	2–8	S	7.03	2500	4	432	584.38	257	1.3	7.5	636	6.33	474	44.5	51	9.7	−1.03	−3.7	−24	
102829	289200	6004650	71–74	S	6.91	2430	2	234	34.16	965	3.1	2.4	336	13.7	694	40.4	95	8.7	−1.08	−5.4	−36	10.3
102830	289200	6004650	16–22	S	6.44	326	3	34	80	33	0.1	0.0	74	0.894	75	3.0	5.5	20.4	−0.71	−6.9	−38	56.2

Lake Cooper

54456	312300	5960200	46–58	C-R	6.53	36800	3	84	39	20300	49	21	3890	78.7	8080	905	3270	17.8	−0.76	−4.6	−32	54.6
54458	310400	5965800	54–62	C-R	6.92	19800	2	174	484	10400	22.5	4	2010	50.3	5380	421	883	8.4	−1.09	−4.7	−33	
57134	305440	5959960	48–66	C-R	6.85	22500	2	144	194	11700	24.5	68	1440	30.9	6150	692	2020	5.8	−1.26	−4.3	−31	3.9
62036	303151	5985250	106–118	C-R	9.88	3460	2	6	501	1710	5.2	1.6	9.3	15.7	1130	9.6	63	0.9	−2.08	−5	−34	39.9

(*continued*)

Table 1. (Continued)

Bore	East[1]	North[1]	Screen[2] m bns	Aquifer[3]	pH	TDS mgL^{-1}	DO$_2$ mgL^{-1}	CO$_2$ mgL^{-1}	HCO$_3$ mgL^{-1}	Cl mgL^{-1}	Br mgL^{-1}	NO$_3$ mgL^{-1}	SO$_4$ mgL^{-1}	K mgL^{-1}	Na mgL^{-1}	Ca mgL^{-1}	Mg mgL^{-1}	Si mgL^{-1}	SiO$_2$(a)[4] SI	δ^{18}O ‰	δ^{2}H ‰	^{14}C pmc
64387	315280	5966587	89–91	C-R	7.36	29000	1	100	123	15800	39	9	2750	68.9	6790	830	2460	3.2	−1.52	−4.2	−31	13.6
73427	313300	5982300	67–69	C-R	7.02	8760	4	46	184	4830	13.4	4.2	741	21.3	2170	393	352	8.4	−1.09	−4.5	−30	18
73537	320700	5980750	50–52	C-R	6.96	12930	3	114	232	6780	19.2	8.8	1010	32.0	3440	386	875	19.9	−0.72	−4.7	−31	53.3
73538	320250	5969750	66–67	C-R	7.73	11710	6	108	340	6080	15.2	8.8	1340	24.2	2690	431	661	6.5	−1.20			15.3
95169	312250	5974400	73–75	C-R	7.18	17900	2	184	293	10300	26	9	1340	25.4	4170	581	1000	2.7	−1.59	−4.8	−34	15.4
95588	314150	5987350	76–78	C-R	7.09	11600	4	70	266	6570	18.8	6.8	929	30.6	2800	299	590	9.5	−1.04	−4.5	−31	21.4
105701	313900	5998950	98–128	C-R	6.41	9280	bd	420	81	5610	14.2	3.2	325	14.5	2280	50.4	447	0.6	−2.25	−4.2	−29	59.1
4868	307525	5940116	35–40	S	7.03	23400	0.1	566	675	12400	20	8	1770	27.8	6030	321	1510	5.2	−1.30	−4.7	−34	18.2
4869	303635	5943212	40–45	S	6.67	5130	1	166	255	2180	4.2	1.2	934	5.2	1450	36.5	90	4.6	−1.35	−5.4	−34	31.6
45458	303635	5943212	20–22	S	6.7	10440	1	406	565	4940	8.4	4.2	1200	12.6	2780	220	285	15.0	−0.84	−4.9	−32	96.7
45459	307525	5940116	2–4	S	7.02	27300	5	688	701	14900	25	97	1880	44.5	5500	850	2570	5.7	−1.26	−4.5	−31	
45460	307525	5940116	11–15	S	6.98	26800	1	754	854	14400	23	32	2070	44.2	5950	254	2410	7.7	−1.13	−4.5	−30	
53671	308150	5955150	37–55	S	6.86	31200	bd	410	263	16600	28	18	3210	24.5	7040	897	2670	7.9	−1.12	−4.1	−30	45
53672	308150	5955150	10–12	S	7.04	26700	4	218	75	14500	25	18	3150	17.0	6030	459	2230	6.4	−1.21	−4.2	−30	
54459	312400	5960150	12–15	S	7.19	20200	4	172	236	11100	26	9	1930	77.2	4940	700	990	5.3	−1.30	−3.8	−26	107.3
57135	305440	5959960	8–10	S	7.37	17290	1	88	262	7870	16	8.8	3600	19.3	3900	474	1040	4.7	−1.34	−4.9	−32	
62037	303200	5985250	12–14	S	9.78	4470	1	8	710	1590	5.2	2.8	739	14.3	1330	5.4	60	1.2	−1.96	−4	−24	120
64388	315281	5966566	3–9	S	7.69	8700	3	130	467	3980	7.2	4.2	1660	15.9	1940	200	290	8.9	−1.07	−3.8	−25	91
73428	313300	5982300	13–17	S	7.03	6050	7	56	128	2930	7.8	15.4	948	11.5	1710	55.8	164	22.6	−0.66	−3.9	−26	
73429	313300	5982300	9–11	S	7.33	4170	2	188	408	1770	5.3	1.5	441	11.3	1210	27.6	78	23.2	−0.65	−3.4	−23	112.2
73539	320250	5969750	16–22	S	6.15	7570	5	114	134	4110	9.8	6.4	731	17.2	1810	188	421	23.9	−0.64	−4.5	−32	
73540	320700	5980750	8–12	S	7.26	3330	6	106	328	1110	5.1	35.6	431	5.5	881	346	65	17.9	−0.76	−5	−30	
80239	316950	5948900	40–46	S	6.27	26900	2	532	298	15100	22	17	2150	57.4	5990	456	2300	4.2	−1.40	−4.7	−32	
95170	312250	5974400	5–8	S	6.63	1848	3	240	355	513	1.15	1.75	191	1.7	496	4.9	28	15.3	−0.83	−2.3	−15	113
95171	312250	5974400	11–14	S	7.15	4160	3	228	412	1650	3.6	8.4	620	4.5	1060	51	107	15.9	−0.81	−3.3	−21	
95590	314150	5987350	5–9	S	7.42	1130	3	210	449	141	0.52	1.76	60	10.7	211	6.5	28	9.7	−1.03	−3.3	−20	109.6
98369	305900	5947450	41–43	S	6.15	29000	2	740	345	16100	25	15	2540	33.1	6100	326	2740	12.4	−0.92	−4.7	−31	
98370	305900	5947450	11–17	S	7.06	4770	2	136	77	2270	3.9	7.2	668	5.0	1430	52	106	7.5	−1.14	−4.6	−32	
98371	311901	5943600	12–24	S	3.52	34400	3	362	bd	18200	27	17	4,690	24.2	7270	750	3010	12.6	−0.92	−4.5	−31	
105702	313900	5998950	4–13	S	6.32	15570	1	480	537	8610	19.6	6.4	963	8.2	3710	373	818	18.5	−0.75	−4.2	−29	94.4
Pyramid Hill																						
51640	232500	5944800	79–119	C-R	8.05	1930	4	124	353	750	1.5	0.2	130	12.1	436	20.1	90	12.1	−0.93	−4.6	−30	
54348	237114	5984181	119–126	C-R	7.47	1040	3	19	1	573	1.2		56	5.7	3 32	18.8	24.4	7.0	−1.17	−5.3	−31	73.4

56029	248932	6037084	124–129	C-R	8.74	10100	bd	46	205	6030	10.2	3.0	527	22.4	2750	10.2	464	6.5	−1.20	−5.2	−34	46.4
60441	253800	5971200	99–102	C-R	8.33	4520	1	6	41	2580	5.2	0.4	231	69.0	1430	33.0	119	7.1	−1.16	−5.0	−34	44.2
79394	234000	6008000	89–92	C-R	7.08	21900	1	144	362	12000	22.0		2450	45.4	5500	446	879	13.9	−0.87	−5.1	−37	
79723	255400	5989000	65–70	C-R	9.57	30500	1	56	60	17200	35.0		2650	20.2	9460	197	841	8.0	−1.11	−5.1	−36	
87806	271900	6015400	156–162	C-R	8.20	4490	2	14	23	2750	5.1	0.1	114	12.1	1420	35.1	110	9.1	−1.06	−5.5	−33	16.6
87807	271900	6015400	122–128	C-R	9.02	4590	2	34	21	2700	5.1	2.8	98	11.5	1560	30.8	115	8.1	−1.11	−5.4	−37	8.2
95040	266700	6001400	83–85	C-R	7.65	7000	2	174	320	2710	5.7	2.0	1300	15.3	2090	151	196	6.6	−1.19	−5.5	−38	6.4
97152	268800	6008950	107–110	C-R	7.76	6220	2	94	217	3140	5.8	0.4	547	10.3	1870	188	138	7.9	−1.12	−5.1	−33	
100503	264800	5980900	130–135	C-R	7.18	13800	1	168	242	6770	13.6		1520	27.6	4480	199	394	7.3	−1.16	−5.1	−35	
108320	235450	5954400	64–70	C-R	7.34	1950	2	147	325	737	1.4	0.2	143	8.1	473	32.3	64	17.2	−0.78	−4.9	−32	
36077	248960	6037040	17–19	S	6.96	16300	6	190	68	8530	14.5	0.5	2190	10.3	4210	324	772	13.0	−0.90	−5.3	−33	66.5
36078	248960	6037040	4–7	S	6.03	28900	2	496	738	14100	26.0	1.0	3850	4.0	7750	510	1400	12.6	−0.92	−5.0	−33	
51718	232500	5944800	42–18	S	6.82	1450	2	90	265	574	1.2	0.1	115	10.0	276	42.2	63.9	16.8	−0.79	−4.7	−31	24.7
51719	232500	5944800	19–23	S	7.23	373	5	94	63	102	0.3	1.2	18	10.1	54	13.2	10.1	6.8	−1.18	−8.5	−53	95.1
51720	233100	5944700	21–24	S	6.95	4790	1	422	558	2070	4.2	3.0	382	21.1	920	195	202	10.9	−0.98	−5.1	−30	
51723	235300	5944800	34–39	S	6.36	1770	7	94	334	707	1.4	0.5	115	16.1	409	40.1	42.6	11.8	−0.94	−4.8	−32	
51724	235300	5944800	14–19	S	7.75	951	6	98	334	190	0.4	5.6	56	3.7	222	11.9	8.5	18.7	−0.74	−5.6	−34	
54350	237114	5984181	35–45	S	9.64	5880	bd	3	140	3390	6.6	0.4	367	20.2	1690	20.2	251	1.1	−1.96	−4.9	−33	24.5
54351	237114	5984181	11–14	S	9.28	13060	2	170	1940	5290	10.5	0.5	1560	31.5	3350	20.2	680	7.1	−1.17	−3.8	−20	109.9
60442	253800	5971300	49–52	S	7.85	9910	1	236	357	5310	11.0	3.0	884	37.7	2260	290	509	11.7	−0.95	−5.2	−34	5.2
60443	253800	5971300	10–16	S	7.06	25200	4	158	597	14200	27.0	18.0	2140	59.0	6490	493	963	15.0	−0.84	−4.7	−30	72.6
68963	268000	5964900	28–32	S	7.66	3770	6	4	107	2000	11.9	5.5	362	11.5	1130	40.1	88	7.1	−1.16	−5.9	−36	
79395	234000	6008000	39–45	S	7.05	47600	2	62	289	26400	47.0	1.0	5810	50.7	12100	909	1870	9.5	−1.04	−4.6	−34	
79396	234000	6008000	4–10	S	7.93	5530	4	122	483	1830	2.6	0.8	1400	6.6	1580	37.0	62.3	9.6	−1.03	−3.7	−25	
79724	255400	5989000	8–11	S	8.69	2310	6	174	365	879	2.0	1.2	198	3.7	631	19.9	29.1	7.0	−1.17	−4.4	−28	
87808	271900	6015400	50–60	S	6.91	19800	bd	210	155	10700	19.5		1920	16.0	6040	207	588	7.3	−1.15	−5.3	−36	12.1
87809	271900	6015400	12–15	S	7.07	43600	2	528	303	23500	41.0	1.3	5130	27.5	11900	685	1460	9.4	−1.04	−4.5	−36	102.6
88238	238400	5965000	25–29	S	7.50	3030	8	18	366	1090	2.3	2.6	412	10.8	990	47.6	80	17.3	−0.78	−5.1	−32	
88239	238400	5965000	8–12	S	7.31	5590	6	116	182	2990	6.2	6.4	478	13.1	1450	122	211	16.8	−0.79	−5.6	−35	
95041	266700	6001400	60–66	S	7.58	9340	4	270	150	5080	10.0	5.7	443	15.6	3020	142	201	7.1	−1.17	−5.3	−37	3.7
95042	266700	6001400	20–26	S	8.53	5020	9	96	320	2380	4.8	15.5	499	15.1	1500	72.8	121	6.1	−1.23	−1.5	−14	103.2
97153	268800	6008950	12–18	S	8.11	24500	6	364	246	13400	25.0	1.7	2030	28.1	7070	479	874	12.2	−0.93	−5.1	−33	
100504	264800	5980900	17–23	S	7.73	1990	12	62	279	784	1.9	0.3	150	9.2	638	21.1	26.1	14.7	−0.85	−5.2	−34	
107928	239300	5943400	40–46	S	5.90	303	4	61	62	87	0.4	1.8	19	6.8	45.5	6.2	5.9	8.3	−1.10			
108319	235450	5954400	35–40	S	7.48	1,760	8	54	195	792	1.7	1.8	133	5.8	465	37.4	61	17.3	−0.78	−4.7	−32	
108321	235450	5954400	11–17	S	7.25	1,660	11	72	204	717	1.8	9.1	141	2.5	433	28.7	34.9	17.0	−0.78	−5.6	−35	

(*continued*)

Table 1. (Continued)

Bore	East[1]	North[1]	Screen[2] m bns	Aquifer[3]	pH	TDS mgL^{-1}	DO$_2$ mgL^{-1}	CO$_2$ mgL^{-1}	HCO$_3$ mgL^{-1}	Cl mgL^{-1}	Br mgL^{-1}	NO$_3$ mgL^{-1}	SO$_4$ mgL^{-1}	K mgL^{-1}	Na mgL^{-1}	Ca mgL^{-1}	Mg mgL^{-1}	Si mgL^{-1}	SiO$_2$(a)[4] SI	δ^{18}O ‰	δ^2H ‰	^{14}C pmc
Ovens																						
11063	439094	5982151	3–5	S	6.66	905			384	201	1.0	10.1	17.1	0.5	253	5	10	23.4	−0.65	−5.5	−36	
11130	449013	6012708	10–13	S	7.66	9,570		180	738	4,150	19.0	2.4	1350	43	2,450	259	363	10.0	−1.02	−4.5	−32	
11134	448902	6011896	12–15	S	7.73	7,100		136	323	2,640	13.0	6.4	1770	29	1,660	242	251	23.5	−0.64	−5.6	−33	
11135	448902	6011897	1–4	S	7.15	4,570			342	1,900	9.2	12.6	609	16	1,410	103	155	10.4	−1.00	−5.0	−30	
11300	432571	5994415	2–4	S	6.34	842			18	471	2.1	0.80	42.8	11	170	66	51	8.4	−1.09	−5.7	−36	
11306	432570	5994419	12–16	S	5.85	749		100	73	342	1.6	0.64	21.4	0.5	103	11	74	21.5	−0.68	−6.0	−37	
11307	432957	5994704	11–15	S	6.52	1,350		104	140	598	2.7	11.5	89	0.5	300	11	69	25.0	−0.62	−5.7	−36	
11308	433687	5995201	10–14	S	6.74	1,230		104	238	456	1.9	12.1	72	0.5	247	11	68	23.4	−0.65	−5.5	−35	
11309	425925	6004547	10–14	S	6.67	1,350		52	342	369	1.5	22.5	139	2	371	9	20	20.9	−0.69	−5.2	−35	
11310	426892	6005377	12–16	S	6.79	1,340		122	250	400	1.6	22.0	113	3	320	32	49	26.3	−0.59	−5.4	−36	
11311	427128	6005376	9–11	S	6.58	1,710		120	262	626	2.4	17.5	136	2	432	21	69	26.0	−0.60	−6.0	−37	
11312	443273	5976526	7–10	S	7.03	911		122	305	153	0.74	6.0	59	0.5	218	6	16	25.8	−0.60	−5.2	−34	
11314	450895	5980002	8–11	S	5.90	687		94	342	114	0.52	3.6	3.0	0.5	98	4	5	22.6	−0.66	−5.0	−34	
11315	451092	5979640	15–21	S	5.79	1,020		100	55	499	1.8	14.2	36.6	0.5	175	48	59	33.8	−0.48	−5.1	−36	
11319	425050	6008216	10–14	S	6.56	1,200		268	177	278	1.2	20.0	117	2	295	4	16	26.4	−0.59	−4.6	−36	
11320	431570	6007088	14–19	S	7.12	7,610		356	1312	2,600	8.8	3.4	860	32	2,200	64	159	21.5	−0.68	−5.6	−37	
11321	439858	6006530	20–24	S	7.22	6,660		254	726	2,900	12.4	0.20	493	18	1,940	157	133	28.6	−0.56	−5.0	−34	
11322	436587	5999712	16–20	S	7.15	10,900		430	1647	3,930	16.0	2.8	1240	41	2,930	216	375	24.1	−0.63	−4.8	−32	
11323	437804	5992767	16–18	S	6.46	449			173	111	0.56	0.01	5.3	0.1	127	3	6	23.3	−0.65			
11324	434880	5990310	11–14	S	6.52	478			93	183	0.88	11.1	25.8	0.1	94	11	39	20.5	−0.70			
11326	439767	5982569	19–24	S	6.60	1,310		186	537	231	1.3	17.8	18.6	2	194	35	64	22.5	−0.66	−5.1	−34	
11328	444131	5966552	6–9	S	6.15	183		64	52	22.7	0.12	0.12	0.3	0.5	15	4	8	16.5	−0.80	−6.6	−40	
11332	435560	5959071	17–22	S	5.11	5,120		160	153	2,520	8.4	bd	587	9	1,240	149	252	40.3	−0.41	−5.1	−33	
11336	455650	5962546	2–8	S	7.81	8,630			3233	2,310	5.2	1.5	474	16	2,470	3	105	7.7	−1.13			
11339	446128	6012282	10–13	S	6.79	12,200		276	836	5,220	18.8	24.0	1656	34	3,270	396	413	25.3	−0.61	−4.3	−32	
11349	455650	5962556	1–4	S	7.67	2,280			988	406	0.61	72.9	67	0.1	704	8	24	9.8	−1.02			
11350	455646	5962544	16–22	S	7.26	1,270			865	66.3	0.20	0.22	0.2	0.1	302	3	31	4.9	−1.33			
11447	447440	5952780	9–12	S	6.64	366			168	73.1	0.40	1.3	1.9	0.1	95	3	7	15.9	−0.81			
48052	481118	5947722	42–45	S	6.29	116			79	12.4	0.02	bd	0.2	0.5	6	5	7	6.2	−1.23	−5.9	−37	
48054	481339	5948034	55–58	S	6.67	331		170	139	1.9	bd	0.01	0.2	0.5	6	3	4	6.2	−1.22	−6.5	−39	
48070	481331	5948002	11–12	S	5.91	118		70	22	2.2	0.13	7.8	1.2	0.5	4	2	3	5.5	−1.28	−6.3	−39	
48071	481114	5947713	5–12	S	5.86	187		88	61	6.0	bd	1.6	6.7	0.5	5	5	7	6.6	−1.19	−5.3	−34	
48073	480788	5947648	4–10	S	6.46	85		42	28	1.2	bd	0.2	0.9	0.5	4	2	2	4.3	−1.39	−5.7	−41	
50788	441959	5998890	60–72	S	7.13	3,220		88	326	1,480	6.2	1.3	275	8	826	92	105	11.4	−0.96			
50789	441956	5998894	18–30	S	6.35	10,200			503	4,970	22.0	2.6	1140	15	2,720	376	413	21.0	−0.69			

1	2	3													
50893	431007	6002593	S	50–52	6.85	355	110	127	26.7	0.1	0.03	0.5	0.1	61	0.1	3	27.0	−0.58	−6.1	−36
51735	498275	5935123	S	30–42	6.85	93	46	33	3.6	0.02	0.36	0.0	0.5	5	2	2	0.9	−2.07	−5.7	−32
51736	498269	5935112	S	20–26	5.65	147	98	29	2.6	bd	0.67	0.2	0.5	8	1	2	5.5	−1.28	−6.4	−36
51737	498330	5935475	S	36–42	6.37	228	132	68	2.0	bd	0.04	0.5	0.5	9	4	6	6.2	−1.23	−5.9	−36
51741	496960	5934606	S	44–50	5.66	132	90	22	1.7	bd	3.3	0.6	0.5	4	1	2	6.8	−1.18	−6.3	−37
51743	499167	5935330	S	5–11	6.06	169	100	45	1.7	0.02	1.9	0.5	0.5	6	3	4	6.7	−1.19	−4.5	−24
51744	498820	5934728	S	6–12	5.37	94	58	17	2.5	0.02	1.8	0.5	0.5	5	1	2	5.1	−1.31	−5.5	−35
51745	499034	5935181	S	5–11	5.42	130	90	24	1.5	bd	1.1	1.3	0.5	3	2	2	6.0	−1.24	−5.2	−33
54981	443094	5984745	S	35–37	7.88	309	62	142	36.7	0.17	bd	0.5	0.5	60	3	4	0.8	−2.12		
62864	434857	5990343	S	111–121	8.95	327	44	139	53.7	0.23	bd	0.2	2	78	3	6	0.5	−2.32		
70027	501416	5979203	S	7–11	6.56	131	58	43	5.8	0.04	3.9	0.1	0.1	9	3	1	10.0	−1.02		
82095	443261	5951687	S	22–28	6.25	346		105	103	0.28	11.9	5.9	1	64	10	24	20.5	−0.70		
83229	474495	5952729	S	8–14	6.14	102	44	39	1.5	bd	2.0	0.3	0.5	3	4	4	3.8	−1.43	−6.5	−37
83232	474773	5953100	S	6–12	5.68	153	80	36	5.8	0.03	2.7	6.1	0.5	8	3	4	6.5	−1.20	−5.3	−33
86160	439858	6006530	S	18–21	10.00	363			217	0.87	0.01	4.2	2	134	4	1	0.2	−2.70	−5.7	−37
88272	493145	5937881	S	10–15	5.91	175	104	47	2.9	0.05	0.29	1.4	0.5	4	5	5	5.2	−1.30	−6.0	−37
88274	490870	5939243	S	35–53	6.00	189	116	45	1.6	bd	0.10	0.6	4	7	2	4	7.9	−1.12	−6.0	−38
93380	504313	5979274	S	56–67	7.66	212	32	122	12.6	0.10	0.05	0.8	3	25	8	8	0.6	−2.28		
93381	504314	5979268	S	12–14	6.90	227	70	116	3.1	0.02	0.03	0.5	2	21	8	6	3.3	−1.51		
93382	503418	5979259	S	5–11	7.07	303	148	122	4.1	0.03	0.04	0.3	0.1	10	4	3	11.0	−0.97		
93383	503820	5979285	S	24–33	7.30	316	74	172	4.3	0.03	0.01	0.9	2	27	19	5	11.3	−0.96		
93384	503817	5979284	S	12–13	7.35	227	106	89	3.8	bd	0.07	0.2	0.1	14	10	3	1.2	−1.95	−5.5	−33
98865	439857	5976832	S	37–39	7.05	227	56	112	3.7	bd	1.6	0.5	0.5	34	2	4	12.9	−0.90	−5.9	−35
98866	439862	5976830	S	18–22	7.49	196	44	88	5.6	0.17	1.5	1.9	0.5	25	2	6	21.4	−0.68	−6.2	−39
98867	439861	5976829	S	5–6	6.34	222	44	90	9.2	0.08	9.7	20.0	5	19	6	11	7.6	−1.14	−5.3	−33
102873	463974	5959641	S	65–69	7.46	159	68	65	2.2	bd	0.08	0.1	0.5	11	5	5	2.4	−1.63	−6.7	−40
109462	497705	5935080	S	45–51	5.70	186	116	43	3.0	bd	0.14	0.2	0.5	6	4	4	9.0	−1.06	−5.8	−37
109652	493158	5937893	S	50–60	5.20	158	108	32	1.6	bd	0.31	1.2	0.5	4	4	2	5.0	−1.32	−5.4	−26
109653	493158	5937893	S	39–42	5.77	192	140	33	1.8	bd	0.20	0.6	0.5	4	2	3	6.5	−1.21	−6.5	−40
110738	444131	5966553	S	44–48	8.72	58		35	7.0	0.03	0.05	0.2	1	8	3	2	1.6	−1.80	−5.5	−31
135123	443255	5951685	S	77–83	7.01	624	140	218	128	0.39	0.02	0.0	3	79	15	23	16.5	−0.80	−5.6	−35
139328	443255	5951685	S	41–49	6.46	179		46	23.6	0.13	0.01	19.9	2	71	3	6	6.9	−1.18	−5.6	−35
302296	437808	5992766	S	71–77	7.71	552	48	220	111	0.49	0.01	24.6	1	122	8	11	6.5	−1.20		

1: eastings and northings relative to the Australian Map Grid (figure 2)
2: depth of screened interval below ground surface
3: B = Basement, C-R = Calivil-Renmark, S = Shepparton
4: Log saturation index relative to aqueous silica (calculated using PHREEQC: Parkhurst and Appelo, 1999)
bd = below detection
Cl/Br ratios, $\delta^{18}O$, δ^2H values for groundwater from all formations in the Ovens, Campaspe, and Lake Cooper areas and ^{14}C contents of groundwater from the Shepparton Formation in those regions area from Cartwright et al. (2006).

this paper (table 1) include major ion, field parameters, and stable and radiogenic isotope data. These were obtained using the methods described by Cartwright and Weaver (2005) and Cartwright et al. (2004, 2006). Briefly, groundwater was sampled from 2002 to 2004 from groundwater monitoring bores that are screened in the Basement, Calivil-Renmark Formation, or Shepparton Formations (table 1), with each bore sampling only one formation. The bores are sealed with a cement/clay seal above the screens to prevent leakage. Bores were sampled using a polyethylene bailer (shallow wells) or a QED Micro Purge bladder pump that was set at the screened interval. Depth to water, pH, alkalinity, dissolved CO_2, dissolved O_2, and temperature were measured immediately in the field. pH, with an uncertainty of <0.1 unit, was measured using an Orion 290 meter and Orion Ross electrodes. Alkalinity and dissolved CO_2 were determined using a Hach digital titrator and reagents; the precision of this technique is better than ±5% of total concentration. Dissolved O_2 was determined using a Hach drop titrator and reagents, a technique that is precise to ±1 mgL^{-1}. Cations were analysed using a Varian Vista ICP-AES at the Australian National University on samples that had been filtered through 0.45 µm cellulose nitrate filters and acidified to pH 2. Anions were analysed on filtered unacidified samples using a Metrohm ion chromatograph at Monash University. Precision of anion and cation concentrations are ±2%. Charge balances calculated using PHREEQC (Parkhurst and Appelo, 1999) were all within ±10%, and most (>80%) were within ±5%. Saturation indices were also calculated using PHREEQC. ^{14}C contents were determined by AMS on the 14UD tandem electrostatic accelerator at the Australian National University. Stable isotope ratios were measured at Monash University using a Finnigan MAT 252 mass spectrometer. $\delta^{18}O$ values of water were measured via equilibration with He-CO_2 at 32°C for 24–48 hours in a Finnigan MAT Gas Bench. δ^2H values of water were measured via reaction with Cr at 850°C using an automated Finnigan MAT H/Device. $\delta^{18}O$ and δ^2H values were measured relative to internal standards that were calibrated using IAEA SMOW, GISP, and SLAP standards. Data were normalised following Coplen (1988) and are expressed relative to V-SMOW where $\delta^{18}O$ and δ^2H values of SLAP are −55.5‰ and −428‰, respectively. Precision (1σ) based on replicate analyses of standards and samples is ±0.15‰ (O) and ±1‰ (H).

3 RESULTS AND DISCUSSION

The data collected in this study combined with that from Calf et al. (1986), Arad and Evans (1987), Macumber (1991), and Cartwright and Weaver (2005) allow a comprehensive understanding of groundwater flow paths and hydrochemical processes in the southern Riverine Province.

3.1 Groundwater flow paths

The pattern of groundwater flow based on groundwater elevations (figure 3) is away from the basin margins along the deep leads. As elsewhere in the Murray Basin (e.g., Herczeg et al., 2001), horizontal hydraulic gradients are low, generally <10^{-3}. Groundwater flow paths extend underneath the River Murray and its major tributaries, indicating that these rivers are not fully penetrating. In the majority of the region, hydraulic gradients within the Shepparton Formation or between the Shepparton and Calivil-Renmark Formations determined from sites with nested bores are downward (figure 3a) with gradients of typically

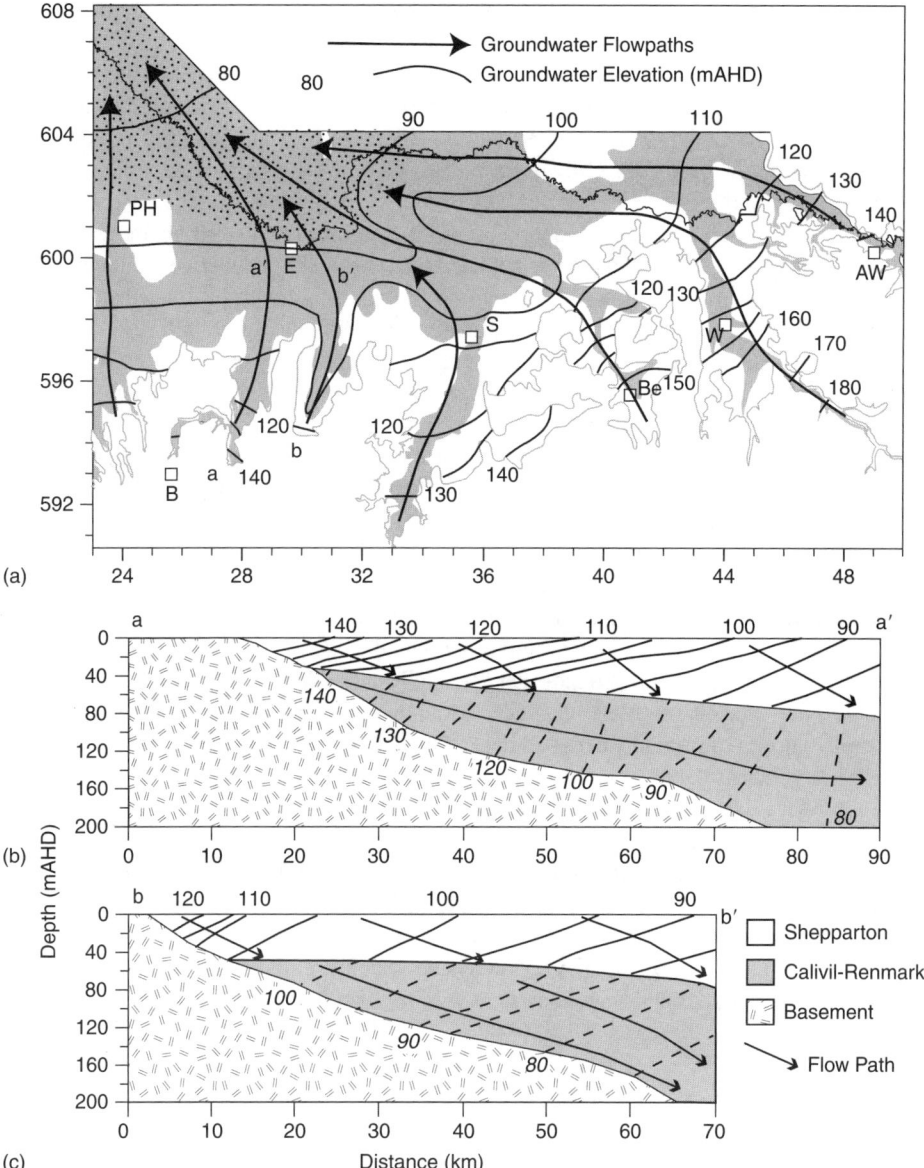

Figure 3. **3a.** Groundwater elevations in the Shepparton Formation above the Australian Height Datum (AHD). The pattern of groundwater elevations in the Calivil-Renmark Formation is similar. Stipple indicates zones of dominantly upwards hydraulic gradients. AW = Albury-Wodonga, B = Bendigo, Be = Benalla, E = Echuca, PH = Pyramid Hill, S = Shepparton, W = Wangaratta. Data from Tickell and Humphries (1985), Dimos et al. (1994), Hennessy et al. (1994). **3b.** Section along the groundwater flow path in the Campaspe Deep Lead (a-a' in figure 3a). **3c.** Section along the groundwater flow path in the Lake Cooper area (b-b' in figure 3a). Sections show the extent of the aquifer units and groundwater elevations (solid lines = Shepparton, dashed = Calivil-Renmark) constructed using data from bores in table 1 and additional bores in the Victorian Water Resources Data Warehouse (http://www.vicwaterdata.net/vicwaterdata/home.aspx).

0.05–0.1 (Cartwright and Weaver, 2005). Within this area of recharge, local (typically 10's to 100's m^2) discharge areas are limited, occurring at the base of steep slopes (e.g. Cartwright et al., 2004) or very locally near rivers. This implies that groundwater recharge can occur broadly across the region and that mixing of groundwater from the Shepparton Formation into the Calivil-Renmark Formation is likely, as indicated by the predominantly downward hydraulic gradients. Only in the northwest of the region are hydraulic gradients upwards (figure 3a). Figures 3b and 3c show sections along groundwater flow paths in the Campaspe (figure 3a) and Lake Cooper (figure 3b) sub-catchments, which are representative of a typical deep lead and the areas between the deep leads, respectively. In both areas, groundwater flow in the Shepparton Formation has a strong downward vertical component. In the Campaspe sub-catchment groundwater flow in the underlying Calivil-Renmark Formation is mainly subhorizontal, while at Lake Cooper flow in the Calivil-Renmark Formation is similar to that in the Shepparton Formation. This reflects the differences in the Calivil-Renmark aquifer between the two regions. In the Campaspe sub-catchment, the Calivil-Renmark Formation contains a higher proportion of sands and gravels than at Lake Cooper, and the contrast in hydraulic conductivity between it and the overlying Shepparton Formation is greater.

Large parts of the southern Riverine Province contain groundwater with total dissolved solids (TDS) contents of >5,000 mgL^{-1} (figure 2). In general, TDS contents of both the Shepparton (figure 2a) and the Calivil-Formation groundwater are similar (figure 2b), with the deep leads containing lower salinity groundwater than the intermediate areas. Deep (>150 m) Calivil-Renmark Formation groundwater generally has TDS <5,000 mgL^{-1} (table 1, figure 6e); however, that is due to this deep groundwater being confined to the deep leads where the majority of groundwater at all depths has low salinity. The distribution of groundwater salinity indicates that groundwater flow paths, especially away from the deep leads, are more complicated than implied by the groundwater elevations. For example, at Lake Cooper groundwater TDS contents in both the Shepparton and Calivil-Renmark Formations decrease along the predicted flow paths (figure 2). Similar declines in TDS contents along apparent flow paths also occur in the Pyramid Hill and Benalla regions. As there is no process other than dilution that can readily lower groundwater salinity, the change in TDS contents is most readily explained by a flow system that has a significant component of vertical as well as subhorizontal flow, consistent with the groundwater flow paths (figs 3b, 3c).

3.2 *Major ions*

Understanding the origins of the solutes in groundwater is required in order to document hydrological processes and to constrain groundwater flow, mixing, and recharge (e.g., Arad and Evans, 1987; Fabryka-Martin et al., 1991; Weaver and Bahr, 1991; Hendry et al., 1991; Herczeg et al., 1991, 2001; Love et al., 1993; Weaver et al., 1995; Radke et al., 2000; Kloppmann et al., 2001; Herczeg and Edmunds, 2000; Dogramaci and Herczeg, 2002; Cartwright et al., 2004, 2006; Cartwright and Weaver, 2005). The occurrence of deep saline groundwater in the southern Riverine Province with ^{14}C ages of up to 28 ka (Calf et al., 1986; Cartwright and Weaver, 2005) indicates that the groundwater chemistry is dominantly the result of processes that operated prior to European settlement and land clearing. Figure 4 summarises major ion data for the southern Riverine Province grouped by sub-catchment and formation. Molar Na/Cl ratios of most groundwater from the Shepparton Formation, Calivil-Renmark Formation, and the basement are 0.6 to 0.8 (figure 4a). Only the least saline groundwater (TDS < 1,500 mgL^{-1}) has significantly higher Na/Cl ratios (up to 20). The higher Na/Cl ratios almost certainly reflect the input of Na from weathering of albitic feldspar (c.f. Herczeg and Edmunds, 2000), which is common in the clastic sediments in the Murray Basin (e.g. Lawrence, 1988; Brown, 1989; Evans and Kellett, 1989; Macumber, 1991). With increasing salinity, the relative proportion of Na decreases and groundwater with TDS > 5,000 mgL^{-1} typically has lower Na/Cl ratios than those of local modern rainfall (~1.2–1.6: Blackburn and McLeod, 1983) or the oceans (0.86). This trend probably reflects the loss of Na by ion exchange with clays in saline environments (e.g. Ghassemi et al., 1995). Trends in the ratios of other major cations (e.g. Ca, figure 4b or Mg, figure 4c), minor cations, and Si to Cl are similar to those of Na/Cl vs Cl (Arad and Evans, 1987; Cartwright and Weaver, 2005), suggesting that rock weathering contributes a relatively high proportion of all solutes to the lowest salinity groundwater. However, Murray Basin groundwater is generally undersaturated with respect to amorphous silica (table 1), implying that rock weathering has not been extensive. Most groundwater also contains measurable dissolved oxygen (up to 12 mgL^{-1}: table 1, Cartwright and Weaver, 2005). Since, progressive mineral dissolution in aquifers generally consumes dissolved oxygen, the near ubiquitous presence of dissolved oxygen is further evidence for limited water-rock interaction.

Low salinity groundwater from the Campaspe, Ovens, and Goulburn deep leads has generally higher cation/Cl ratios than groundwater from the intermediate areas. This probably reflects the differing nature of the Shepparton Formation sediments that in the deep leads are coarser grained and less mature, with higher volumes of weatherable minerals such as feldspars (Lawrence, 1988; Brown, 1989; Evans and Kellett, 1989; Macumber, 1991).

There is no correlation between Ca or Ca + Mg and total dissolved inorganic carbon (table 1, Arad and Evans, 1987; Hannam et al., 2004; Cartwright and Weaver, 2005), implying that calcite or dolomite dissolution does not control the geochemistry of the Riverine Province groundwater. This is consistent with the low carbonate content of the aquifers in this part of the Murray Basin. The lack of correlation between Ca and S likewise implies that gypsum dissolution does not control groundwater chemistry. Gypsum dissolution is a likely source of S. However, Ca/S ratios are generally <1, especially in the Shepparton Formation groundwater (table 1, Cartwright and Weaver, 2005). Carbonate precipitation in the soils during recharge or the formation of clay minerals during weathering may explain the low Ca/S ratios. Alternatively, much of the S may be derived from evaporation of rainfall, which locally has a Ca/S ratio of 0.6–0.8 (Blackburn and McLeod, 1983), similar to that of many of the samples. Some lower salinity groundwater samples have Ca/S ratios >1 that again probably reflect silicate weathering. The most saline groundwater is close to saturation with respect to gypsum (e.g. Cartwright and Weaver, 2005; Hannam et al., 2004), indicating that gypsum precipitation may limit S concentrations of those samples.

Molar Cl/Br ratios range from 50 to 1,600, and while they are most variable at low salinities (TDS < 1,500 mgL^{-1}), they are largely invariant with increasing salinity. The lowest Cl/Br ratios are interpreted to reflect those of rainfall in the area. While there are no systematic measurements of Cl/Br ratios in rainfall from southeast Australia, surface water samples from the Ovens and Goulburn Rivers (figure 2) have molar Cl/Br ratios of 180–220 (figure 4), similar to the lowest Cl/Br ratios in groundwater. Additionally, in semi-arid continental interiors, groundwater with Cl/Br ratios lower than the oceans is common (e.g., Fabryka-Martin et al., 1991; Herczeg et al., 1991; Davis et al., 1998, 2001; Harrington and Herczeg, 2003; Cartwright et al., 2004, 2006). Groundwater with the highest Cl/Br ratios has most probably

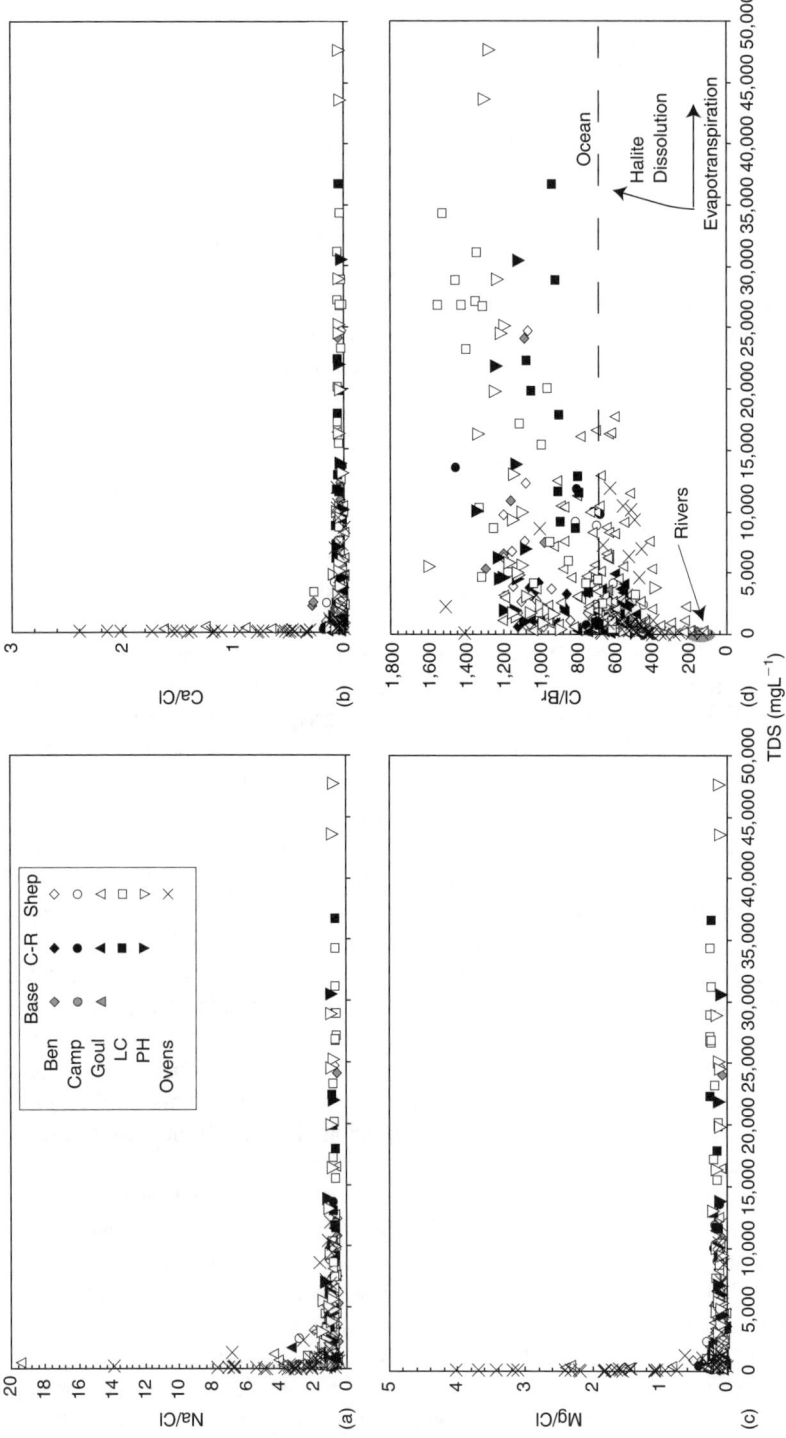

Figure 4. Summary of geochemistry of the southern Riverine Province groundwater. **4a.** Molar Na/Cl vs TDS (mgL^{-1}). **4b.** Molar Ca/Cl vs TDS (mgL^{-1}). **4c.** Molar Mg/Cl vs TDS (mgL^{-1}). **4d.** Molar Cl/Br vs TDS (mgL^{-1}), ocean value indicated by dashed line. Subcatchments: Ben = Benalla, Camp = Campapse, Goul = Goulburn, LC = Lake Cooper, PH = Pyramid Hill. Aquifers: Base = basement, C-R = Calivil-Renmark, Shep = Shepparton. The data imply that evapotranspiration with minor silicate, halite, and carbonate dissolution is the dominant hydrochemical process. Except for Cl/Br ratios, there is no difference in the groundwater chemistry between aquifers or areas. Data from table 1, Cartwright and Weaver (2005), Arad and Evans (1987), Macumber (1991).

dissolved small volumes of halite. However, halite in the Riverine Province has Cl/Br ratios of $\sim 10^4$ (Cartwright *et al.*, 2004), similar to that elsewhere (McCaffrey *et al.*, 1987; Kloppmann *et al.*, 2001), hence the amount of halite dissolution must be minor. By mass balance, if the water that recharges the system has approximately the Cl concentration of modern rainfall ($\sim$0.05 mmol/L: Blackburn and McLeod, 1983), and a Cl/Br ratio of 200, then dissolution of only 1 mmol of halite per litre of water would raise Cl/Br ratios to >1,600, which are higher than those recorded in any groundwater from the southern Riverine Province. The relative invariance of Cl/Br ratios with increasing salinity confirms that evapotranspiration is the dominant mechanism in increasing salinity of the groundwater. The dominance of evapotranspiration implies that the relative differences in salinity between subcatchments reflect recharge rates. In the deep leads, relatively higher rates of recharge through coarser-grained sediments produces fresh groundwater, while slower recharge rates through the more clay-rich soils and lower hydraulic conductivity Shepparton Formation sediments outside the deep leads results in higher groundwater salinities.

Average Cl/Br ratios in the deep leads are significantly lower (Campaspe $\sim$705, Goulburn $\sim$675, Ovens $\sim$550) than those in the intermediate areas (Pyramid Hill $\sim$1,165, Benalla $\sim$1,100, Lake Cooper $\sim$905). Cartwright *et al.* (2006) attributed this difference to the groundwater in the intermediate areas dissolving slightly more halite during recharge than that in the deep leads. That study further proposed that these small volumes of halite were windblown from central Australia and/or deposited by evaporation in dry summers.

These data imply that the dominant hydrochemical process throughout the southern Riverine Province is evapotranspiration during recharge with minor local silicate weathering, halite dissolution, carbonate dissolution and re-precipitation, ion exchange, and reactions between clay minerals. The dominant processes do not vary between the deep leads and the adjacent more saline areas or between the Shepparton Formation and Calivil-Renmark Formation. There is also no correlation of groundwater chemistry (e.g., TDS contents, ionic ratios, or saturation indices) with groundwater age or position along flow paths, suggesting that the hydrochemical processes have operated in a similar manner for several thousand years. Evapotranspiration is also the dominant hydrochemical process in the Mallee-Limestone Province (Allison *et al.*, 1990; Love *et al.*, 1993; Herczeg *et al.*, 2001; Dogramaci and Herczeg, 2002), and probably throughout the Murray Basin.

3.3 *Oxygen and hydrogen isotopes*

Figure 5a shows $\delta^{18}O$ and δ^2H values of southern Riverine Province groundwater. Groundwater from all subcatchments clusters around the global and Melbourne meteoric water lines at approximately the composition of modern precipitation for Melbourne ($\delta^{18}O = -5.0$‰, $\delta^2H = -28$‰). The occurrence of samples to the left of the Melbourne meteoric water line is probably due to local climatic differences between Melbourne (which is on the coast) and the southern Riverine Province (which is inland and more arid) resulting in displacement of the local meteoric water line to the left of the Melbourne meteoric water line. A similar shift of $\delta^{18}O$ and δ^2H values of groundwater from Cobram in Northern Victoria, which is also inland, has also been recorded (Ivkovic *et al.*, 1998). The data as a whole defines an array with a slope of $\sim$5, suggesting that the stable isotopes reflect the effects of evaporation. However, most samples show an increase in $\delta^{18}O$ of <3‰ and there is no correlation of $\delta^{18}O$ values with TDS (figure 5b). The data of Gonfiantini (1986) together with evaporation experiments in the laboratory (Cartwright, unpublished data) suggest that a $\sim$5‰ increase in $\delta^{18}O$ values is produced by 20% evaporation, which is far less than that required to produce the

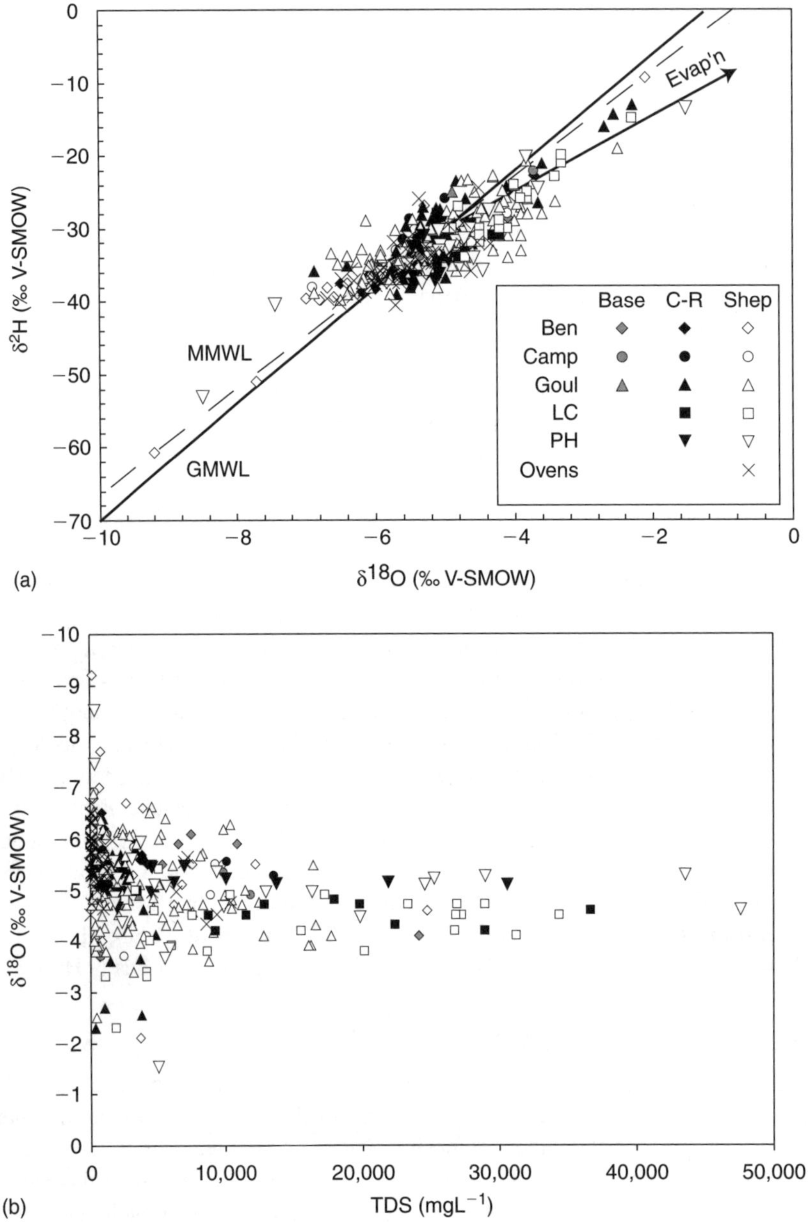

Figure 5. **5a.** $\delta^{18}O$ vs δ^2H values for southern Riverine Province groundwater. Data cluster around the global (GMWL) and Melbourne (MMWL) meteoric water lines at about the value of modern rainfall in Melbourne (Cartwright, unpubl. data). Symbols as for figure 4. The arrowed line is a linear best fit to the entire dataset. **5b.** $\delta^{18}O$ vs TDS values for southern Riverine Province groundwater. The lack of a positive correlation suggests that transpiration rather than evaporation is the more important process in controlling groundwater salinity. Data from table 1, Cartwright and Weaver (2005), Macumber (1991).

high TDS contents. Transpiration, which does not significantly affect $\delta^{18}O$ values (Clark and Fritz 1997), may be the more important process. Until recent land clearing, the native vegetation in southeast Australia was an efficient user of available rainfall leading to significant transpiration. Alternatively, evaporation from the water table may occur in an environment with a thicker vapour boundary layer and a higher humidity (c.f., Herczeg et al., 1992). Both these factors potentially reduce the enrichment in ^{18}O and ^{2}H over that resulting from evaporation from open water, from which the experimental fractionations are derived (Gonfiantini, 1986). Thus, the degree of evaporation may be higher than predicted from the stable isotope data. Unlike the groundwater from the Mallee-Limestone Province (Herczeg et al., 2001), $\delta^{18}O$ values do not vary with distance from the basin margins, and no regions that contain low-$\delta^{18}O$ value palaeowaters (c.f. Leaney and Herczeg 1999; Leaney et al., 2003) exist.

3.4 Variations in chemistry as indicators of groundwater flow

In addition to understanding the sources of solutes, variations in groundwater chemistry are potentially important qualitative indicators of groundwater flow paths. In individual subcatchments Cl/Br ratios and $\delta^{18}O$ values become more homogenous with depth (figure 6a,b). The tendency for $\delta^{18}O$ values to be homogenised along groundwater flow paths is well known (e.g., Clark and Fritz, 1997; McGuire et al., 2002; Goller et al., 2005), and the trend in $\delta^{18}O$ values is consistent with groundwater in the Shepparton Formation having a major component of vertical flow which would have homogenised groundwater with different initial $\delta^{18}O$ values. A similar process of homogenisation occurs with Cl/Br ratios. Mixing during vertical flow homogenises the Cl/Br ratios of groundwater that has dissolved differing, albeit minor, quantities of halite during recharge or was derived from rainfall events that had different Cl/Br ratios.

Such homogenisation may be expected in $\delta^{18}O$ values and conservative ions such as Cl and Br that are dominantly derived from rainfall. However, in all subcatchments except for the Ovens Valley, the cation/Cl ratios also show less variability with depth. Aside from the Ovens Valley groundwater, groundwater with Na/Cl ratios >2 occurs mainly at depths of <20 m (figure 6c). This is not a function of the salinity distribution as the shallow groundwater generally is at least as saline as the deeper groundwater (figure 6d). These data imply that the influence of rock weathering is observed mainly in the shallow groundwater. There are several reasons why this may be so. Firstly, the shallowest sediments may contain higher concentrations of reactive minerals such as feldspars; however, the reported sedimentology (Lawrence, 1988; Brown, 1989; Evans and Kellett, 1989; Macumber, 1991) does not support this. Alternatively, the rise of the water table following land clearing has caused the shallower parts of the aquifer that were previously in the unsaturated zone to now be part of the saturated zone. However, this would only result in shallow groundwater having high Na/Cl ratios if water-rock interaction were restricted to the unsaturated zone. Mineral dissolution is probably more common in the unsaturated zone where CO_2 and O_2 concentrations are higher (e.g. Drever, 1997). It is more likely that as the high cation/Cl ratios are recorded in only some of the shallow groundwater samples, mixing during vertical flow homogenises cation/Cl ratios in a similar way to the homogenisation of Cl/Br ratios or $\delta^{18}O$ values and that the initial variable cation/Cl ratios reflect changes in local mineralogy and extent of weathering.

Nitrate concentrations are also qualitative tracers of recent groundwater flow. High (up to 5.2 mmolL^{-1}) nitrate concentrations in shallow groundwater from the Shepparton Formation

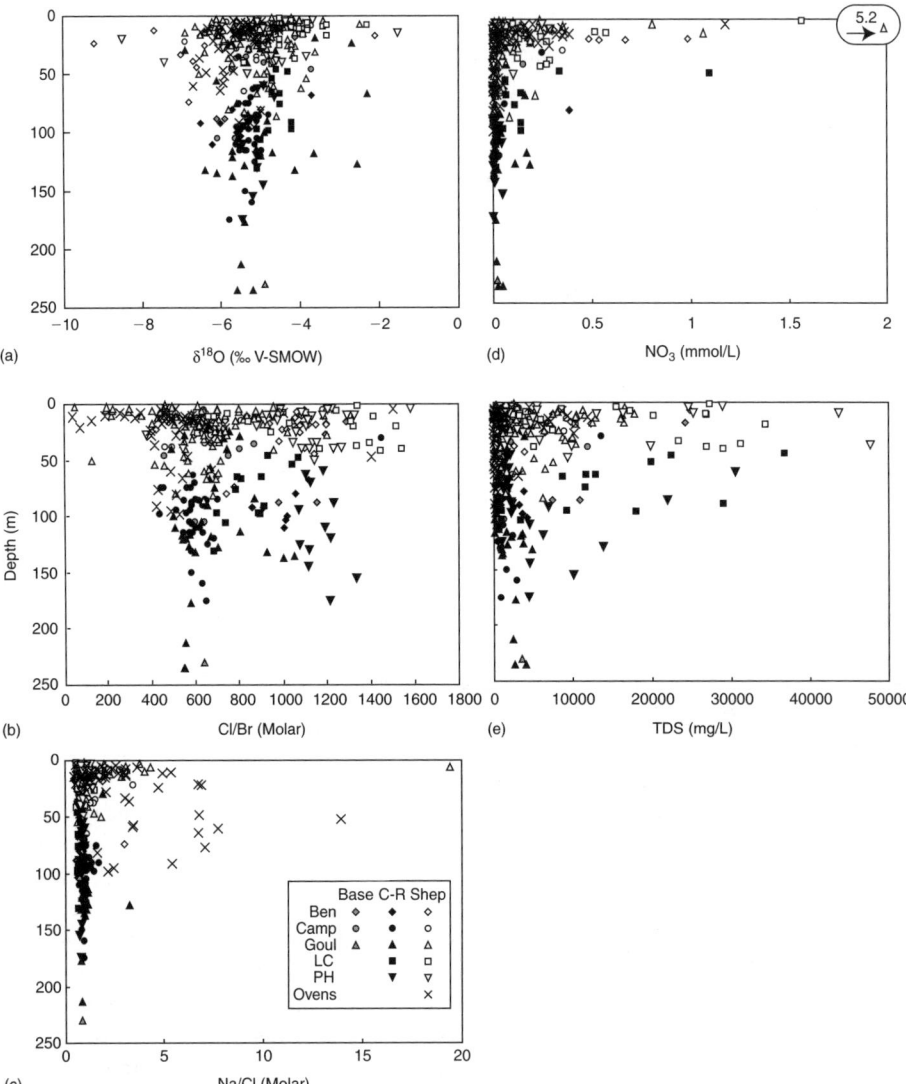

Figure 6. Variation of $\delta^{18}O$ values (**6a**), Cl/Br ratios (**6b**), Na/Cl ratios (**6c**), NO_3 concentrations (**6d**), and TDS contents (**6e**) with depth. Symbols as for figure 4. Within each subcatchment, groundwater chemistry becomes more uniform with depth implying mixing during vertical flow. The difference in Cl/Br ratios between the different subcatchments is clearly seen. High nitrate concentrations in the shallow groundwater are probably due to modern agricultural practices. Data from table 1, Cartwright and Weaver (2005), Arad and Evans (1987), Macumber (1991).

likely represent contamination from modern agricultural practices in the region. Nitrate concentrations in deeper (>20 m) groundwater are generally <0.4 mmolL^{-1} (figure 6e), reflecting the fact that the deeper groundwater was recharged largely prior to the establishment of modern agriculture. Locally in the Goulburn Valley deep lead, groundwater from the Calivil-Renmark Formation at ~120 m has high nitrate concentrations (figure 6e) and anomalously high pmc contents (Cartwright and Weaver, 2005), suggesting recent local leakage of shallow groundwater or surface water to depth.

In some of the subcatchments (e.g. the Goulburn Valley and the Campaspe deep leads), there are general increases in salinity along flow paths in the Calivil-Renmark Formation (figure 2b). In both cases there is a similar increase in salinity in groundwater in the overlying Shepparton Formation (figure 2a). Understanding whether the increase in salinity in the deeper aquifers is due to progressive water-rock interaction, leakage of groundwater from the overlying Shepparton Formation, or represents pulses of water recharged under different climatic conditions is important in constraining groundwater flow paths. Neither Cl/Br nor Na/Cl ratios increase along these flow paths (figure 7a,b) precluding progressive dissolution of halite or silicate minerals as the cause of the increasing salinity. Herczeg *et al.* (2001) suggested that lateral variations in salinity in deep groundwater from the Mallee-Limestone province might reflect long-term (thousand year timescale) changes in rainfall and evapotranspiration rates. However, the lateral trends in salinity in the Mallee-Limestone Province are accompanied by changes in stable isotope ratios (Herczeg *et al.*, 2001) that also plausibly reflect climate change. This is not the case in the Goulburn and Campaspe deep leads (figure 7c). Additionally, the Renmark Group aquifer in the Mallee-Limestone Province is separated from shallower aquifers by low conductivity clays and marls, whereas such units are generally absent in the southern Riverine Province (figure 2). This makes inter-aquifer mixing more likely in the southern Riverine Province than in the Mallee-Limestone Province. It is most likely that changes in salinity in the deeper aquifers in the southern Riverine Province dominantly reflect the leakage of water from the overlying Shepparton Formation. The lateral change in groundwater chemistry at Lake Cooper, outside the deep leads, also implies considerable vertical flow and mixing. Groundwater salinity in this sub-catchment in both the Shepparton and Calivil-Renmark Formations decreases northward along the groundwater flow path (figure 2). This change is accompanied by a decrease in Cl/Br ratios (figure 7d). Below salinities where halite precipitation occurs, there are no hydrochemical processes than can reduce groundwater salinity or decrease Cl/Br ratios and the only explanation is that these trends are due to the vertical mixing of relatively fresh shallow groundwater in the north of the catchment with more saline laterally-flowing groundwater (c.f., figure 3c).

3.5 *Radiocarbon*

Pmc contents of the southern Riverine Province groundwater confirm many of the conclusions regarding groundwater flow made on the basis of the major ion chemistry, and allow quantification of the timescales of groundwater flow. In the Shepparton Formation, there is little correlation of groundwater pmc values with distance from the basin margins (figure 8a); however, within each subcatchment, pmc values decrease with depth (figure 8b). This implies that flow within the Shepparton Formation throughout the southern Riverine Province has a strong downward vertical component. Correction of ages for the dissolution of "dead" carbon from the aquifer matrix is not straightforward. For example, Calf *et al.* (1986) used a $\delta^{13}C$ correction assuming closed-system congruent carbonate dissolution (c.f. Clarke and Fritz, 1997). However, $\delta^{13}C$ values in that study ranged between −18.5 and −4.8‰. Assuming $\delta^{13}C$ values of soil zone and matrix carbon of −25‰ and 0‰, respectively (Clarke and Fritz, 1997), a pH of 7, and a temperature of 25°C, the calculated contribution of matrix carbon to the total DIC ranges from 22 to 80%. However, as

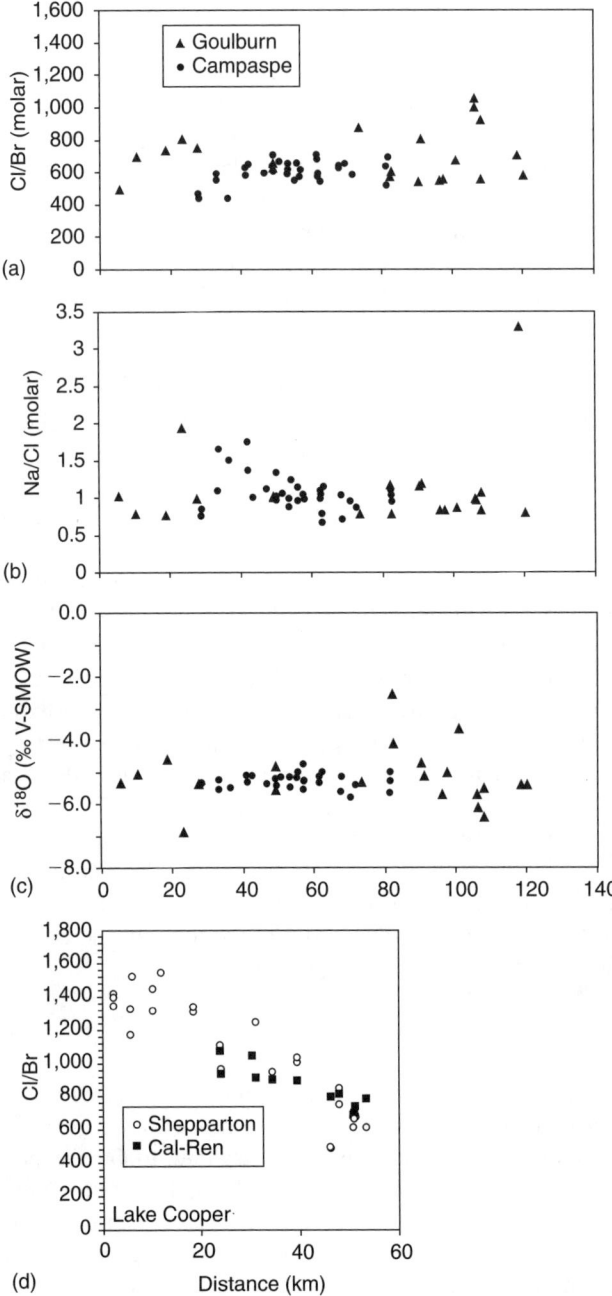

Figure 7. Variations in Cl/Br ratios (**7a**), Na/Cl ratios (**7b**) and $\delta^{18}O$ values (**7c**) in Calivil-Renmark groundwater with distance along the Goulburn and Campaspe deep leads (from the southern end of the lead). Despite the relatively regular increase in groundwater ages along these deep leads, there is little difference in groundwater chemistry. **7d.** Variation in Cl/Br ratios of groundwater from the Calivil-Renmark and Shepparton Formations with distance away from the basin margins in the Lake Cooper region. Data from table 1, Cartwright and Weaver (2005), Arad and Evans (1987).

discussed above and by Arad and Evans (1987), Macumber (1991), and Cartwright and Weaver (2005) groundwater chemistry and the general lack of carbonates in the aquifers (Tickell, 1978; Tickell and Humphries, 1985, 1986; Lawrence 1988; Macumber, 1991) make it unlikely that there is significant carbonate dissolution. The origins of the widely variable $\delta^{13}C$ values of DIC reported by Calf et al. (1986) and Cartwright and Weaver (2005) are not well understood and using them to correct the ^{14}C data is consequently problematic. The stable isotope correction also requires assumptions to be made about the $\delta^{13}C$ values of the soil CO_2 and the pH during recharge that are not well constrained. Alkalinity corrections (c.f. Clarke and Fritz, 1997) also yield variable results and Calf et al. (1986) do not present the data required for that correction. Given these problems, a simple correction has been employed. As groundwater geochemistry implies that only a small percentage of carbon is likely to be derived from carbonate minerals such as calcite cements, a contribution of 15% matrix carbon is used to correct the pmc values (c.f., Vogel, 1970). While this is a simple correction, given the relative uniformity of the aquifer mineralogy and groundwater chemistry across the southern Riverine Province, the relative ages of the samples to each other is likely to be correct. Using this correction, groundwater from the base of the Shepparton Formation in the areas where it is thickest is up to 26 ka.

Shepparton Formation groundwater from <20 m depth commonly yields modern ages, implying that it contains water recharged during the atmospheric nuclear tests in the 1950's and 1960's. Water table depths in the southern Riverine Province have risen by up to 20 to 30 m over the last 200 years since land clearing (e.g. Ghassemi et al., 1995), and the aquifer intervals sampled by many of the shallower bores would have been in the unsaturated zone prior to that time. Groundwater in the Shepparton Formation from the Goulburn and Campaspe deep leads shows similar trends of decreasing pmc values with depth (figure 8a) that are steeper than the pmc vs depth trends from the Pyramid Hill and Lake Cooper regions. This implies that vertical infiltration rates in the deep leads are higher than in those intermediate areas, which as discussed above is consistent with the salinity differences between the deep leads and adjacent areas. Paradoxically, the pmc vs depth trend for the relatively-saline Benalla region is steeper than those in the deep leads; however, it is defined by only one deep sample that is in a coarse-grained unit near the base of the Shepparton Formation. This part of the Shepparton Formation contains low salinity groundwater that was probably rapidly recharged. The scatter in the pmc vs depth trends in each area reflects the variability in vertical infiltration rates due to the heterogeneous hydraulic conductivity of the Shepparton Formation.

Vertical flow rates and hydraulic conductivities were estimated for the Shepparton Formation based on the pmc trends. Pmc vs depth trends for the combined deep lead and intermediate areas, with the exception of the one deep Shepparton Formation sample from the Benalla area, have r^2 values of 0.7–0.8. The general trends of age with depth imply infiltration rates of approximately 4–5 mm/year for the deep leads and 1–2 mm/year in the intermediate areas. For porosities of 0.1–0.3, these equate to recharge rates of 0.4–1.5 mm/y to 0.1–0.6 mm/year in the deep leads and intermediate areas, respectively ($\leqslant$1% of modern rainfall). These are similar to recharge rates estimated in the Mallee-Limestone Province of the Murray Basin by Allison et al. (1990). Similar estimates for pre-land clearing recharge rates in the southern Riverine Province are obtained by Cl mass balance. For example, Cl concentrations of groundwater in the Lake Cooper area are typically ~3,000 to 20,000 mgL^{-1} (figure 2; table 1). Annual rainfall in this area is 450–475 mm

(Bureau of Meteorology, 2005) and rainfall in the southeast Murray Basin contains $\sim 1.5\,\mathrm{mgL}^{-1}$ Cl (Blackburn and McLeod, 1983). Using the relationship:

$$R = P * Cl(p)/Cl(gw),$$

where R is recharge, P is annual rainfall, and Cl(p) and Cl(gw) are the Cl concentrations in rainfall and groundwater, respectively (e.g., Allison et al., 1990) yields recharge estimates of 0.03 to 0.2 mm/yr. Estimated recharge rates in the Campaspe and Goulburn Valley deep leads where groundwater typically has Cl concentrations of 500–2,000 mgL^{-1} (Cartwright and Weaver, 2005; table 1; figure 2) and where annual precipitation is 550–600 mm (Bureau of Meteorology, 2005) are 0.4 to 1.8 mm/year. A 20 m rise in the water table over the last 200 years following land clearing (Ghassemi et al., 1995) would require that recharge rates had increased to 10–30 mm/year ($\sim$2–8% of modern rainfall).

For an average vertical hydraulic gradient of 0.05–0.1 and porosities of 0.1 to 0.3, vertical hydraulic conductivities calculated using Darcy's Law are approximately 3×10^{-5} m/day in the deep leads and 1×10^{-5} m/day in the surrounding areas. That these are lower than vertical hydraulic conductivities for the Shepparton Formation reported by Tickell (1978, 1991), Tickell and Humphries (1986), and Arad and Evans (1987) based on pump tests ($10^{-5}-10^{-1}$ m/day) is probably due to a variety of factors. Firstly, they were calculated using present day hydraulic gradients that, due to the recent water table rise following land clearing will be higher than historical ones. Additionally, these values assume that recharge occurs through a fully saturated system. The hydraulic conductivity of unsaturated sediments in semi-arid regions is significantly lower than that of the equivalent saturated sediment (e.g. Ragab and Cooper, 1993). Thus, the calculated hydraulic conductivities will be minima for the saturated zone. This will especially be the case, as prior to land clearing, the unsaturated zone in this region would have been substantially thicker than at present. Thus, the hydraulic conductivities estimated from the ^{14}C data will be consistently lower than those estimated from pump tests.

Interpretations of the pmc data in the Calivil-Renmark Formation groundwater are more problematic due to the leakage through the Shepparton Formation. Calf et al. (1986) used part of the data presented in figure 8 to contour groundwater ages and transmissivities within the Calivil-Renmark Formation. However, the expanded data set shows that the distribution of pmc values within the Calivil-Renmark Formation is highly irregular, probably due to the variable leakage rates through the Shepparton Formation combined with variable flow paths controlled by heterogeneity within the Calivil-Renmark Formation. Only in the Campaspe deep lead and, to a lesser extent in the Goulburn Valley deep lead, is there a regular decrease in pmc contents along the groundwater flow paths constructed using the groundwater elevations (figure 3). This suggests that groundwater flow in the Calivil-Renmark Formation in these areas has a larger component of lateral flow compared to the rate of vertical leakage. By contrast, away from these deep leads, the heterogenous distribution of pmc values implies that the ratio of lateral flow to vertical flow in these areas is lower, which as discussed above is also implied by the heterogeneous distribution of salinity. The scatter of pmc values in the Calivil-Renmark Formation is such that it is extremely difficult to use these data to define aquifer properties with any confidence.

In general, groundwater in the areas between the deep leads is older than in the deep leads themselves. This is due to the lower rates of vertical flow through the Shepparton Formation away from the deep leads (discussed above) combined with slower lateral flow

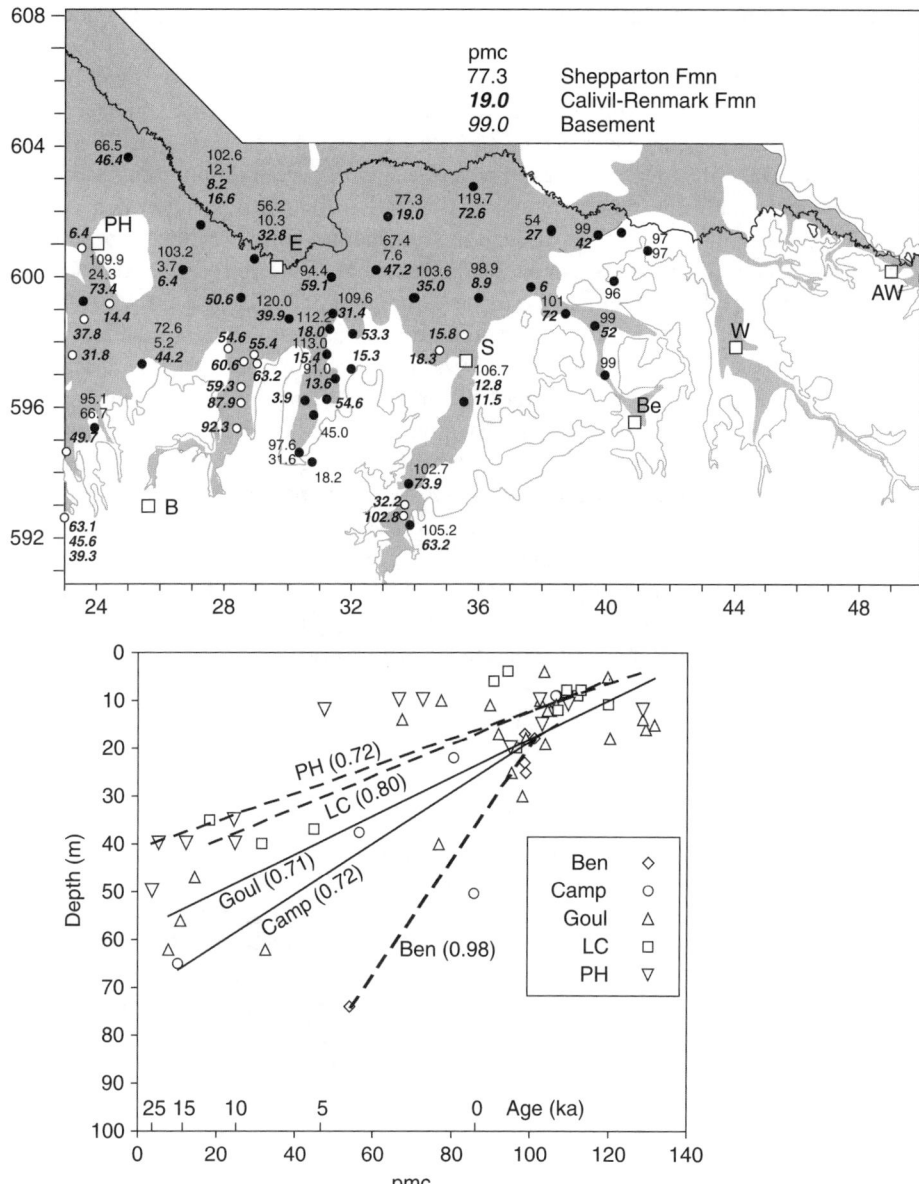

Figure 8. **a.** Pmc contents of groundwater from the Shepparton Formation, Calivil-Renmark Formation, and basement across the southern Riverine Province. Data from table 1, Cartwright and Weaver (2005), and Calf *et al.* (1986: open circles). AW = Albury-Wodonga, B = Bendigo, Be = Benalla, E = Echuca, PH = Pyramid Hill, S = Shepparton, W = Wangaratta. **8b.** Variation of pmc values with depth for Shepparton Formation groundwater. Lines are best fit to the datasets from the different subcatchments, numbers in brackets are the r^2 values. Ages are calculated assuming 15% dilution by matrix carbon. Symbols as for figure 4. Data from table 1, Cartwright and Weaver (2005), Calf *et al.* (1986).

in the deeper aquifers that in those areas contain finer-grained sediments. Groundwater from the Calivil-Renmark with the lowest pmc contents is from the margins of the basin away from the deep leads in areas such as Lake Cooper (pmc as low as 3.9) and Benalla (pmc as low as 6). Groundwater in these areas probably recharges slowly and lateral flow rates are limited. The groundwater in both of these areas is highly saline (figure 2), which is consistent with the inference that rates of recharge control overall groundwater salinities.

Fully understanding the pmc contents of the Calivil-Renmark groundwater requires detailed modelling of vertical flow through the Shepparton Formation that is beyond the scope of this study. Nevertheless, some broad constraints on the age of groundwater from the Calivil-Renmark Formation may be made. Groundwater ages in the Calivil-Renmark Formation are up to 25 ka with the oldest groundwater commonly close to the basin margins in areas of high salinity (e.g. at Lake Cooper). In the north of the Campaspe and Goulburn deep leads where groundwater feeds into the Murray Basin deep lead, groundwater is up to 8 and 20 ka, respectively. As groundwater at the base of the Shepparton Formation commonly has lower pmc contents than that in the underlying Calivil-Renmark Formation (figure 8) the leakage through the Shepparton Formation increases the average age of the deep groundwater indicating that these ages are maximum estimates. By contrast, in some areas, such as the Goulburn Valley, groundwater with anomalously young ages and high NO_3 concentrations is locally present in the Calivil-Renmark Formation (Cartwright and Weaver, 2005), indicating that, locally, rapid leakage probably through interconnected sand lenses in the Shepparton Formation also occurs.

4 CONCLUSIONS

The combination of groundwater elevations and groundwater geochemistry allows regional flow systems to be constrained. The dominant processes controlling the hydrochemistry of groundwater in the southern Riverine Province of the Murray Basin is evapotranspiration with only minor silicate, halite, carbonate, and gypsum dissolution. The overall salinity of the groundwater is, therefore, controlled by recharge rates. The pmc values indicate that the lower salinity deep leads are areas of more rapid recharge, probably due to the Shepparton Formation in those areas containing coarser-grained sediments. Based largely on data from the Mallee Limestone Province, Herczeg et al. (2001) concluded that groundwater in the Murray Basin had attained close to its final chemical composition resulting from processes in the unsaturated zone. The data from the southern Riverine Province presented here confirm that progressive water-rock interaction during groundwater flow is limited and that any changes to groundwater chemistry subsequent to recharge reflect mixing of groundwater from different aquifers.

The distribution of salinity, trends in major ion and stable isotope ratios, and the distribution of pmc values imply that groundwater flow in the southern Riverine Province is locally more complex that may be concluded from the groundwater elevations alone. Throughout most of the southern Riverine Province, vertical flow occurs within the Shepparton Formation while the Calivil-Renmark Formation shows a greater component of lateral flow. However, there are differences between the subcatchments. In the deep leads, the distribution of groundwater elevations and pmc values implies that there is significant lateral flow in the deeper Calivil-Renmark Formation. By contrast, in the intermediate areas, the groundwater elevations, variations in salinity, and distribution of pmc values imply much greater relative vertical leakage through the Shepparton Formation into the Calivil-Renmark Formation. Within the deep leads, there are variations in the degree of leakage through the Shepparton Formation. For example, salinity values and pmc contents of groundwater from the Calivil-Renmark Formation in the Goulburn Valley deep lead are more variable than those from the Campaspe deep lead, suggesting a greater component of vertical leakage from the Shepparton Formation in the Goulburn Valley. In the Lake Cooper region, the decrease in salinity and Cl/Br ratios away from the basin margins implies considerable dilution of deeper groundwater by relatively fresh shallow groundwater along the flow path.

Groundwater from the Calivil-Renmark in the Goulburn Valley deep lead locally has high NO_3 concentrations and pmc contents, suggesting that recent leakage of surface water or shallow groundwater into the deeper aquifers in the Goulburn Valley deep lead locally occurs. Similarly, groundwater in the Ovens deep lead has very heterogeneous chemistry with depth, suggesting that rapid vertical flow with little mixing occurs. The likelihood of short- or longterm vertical flow through the Shepparton Formation occurring is probably controlled by the degree to which sand lenses in the Shepparton Formation are vertically interconnected. The deep groundwater in many of the deep leads is a viable resource and using geochemistry to identify where potential surface contamination may occur is valuable in protecting that resource.

Recent land clearing has dramatically increased recharge in the southern Murray Basin from typically 0.05–0.1 mm/a (~0.1% of annual rainfall) to 1–50 mm/a, or up to 10% annual rainfall (e.g., Allison et al., 1990 and Ghassemi et al., 1995). This has caused the water table to rise and brought saline groundwater closer to the land surface resulting in the well-documented dryland salinity problem. The rise in the water table increases both lateral and horizontal hydraulic gradients that will promote flow between the Shepparton and the Calivil-Renmark Formations, which may have a deleterious impact on the future quality of the deeper groundwater.

ACKNOWLEDGEMENT

Marcus Onken, Kaye Hannam, and Margaret Lauricella helped collect the samples that were analysed by Ben Petrides (anions and stable isotopes), Andy Christy (cations), and Fred Leaney (^{14}C). Draga Geldt drafted figure 2. The manuscript benefited from helpful reviews from J Carrillo and an anonymous reviewer.

REFERENCES

Allison GB, Cook PG, Barnett SR, Walker GR, Jolly ID, Hughes MW (1990) Land clearance and river salinisation in the western Murray Basin, Australia. J Hydrol 119: 1–20

Arad A, Evans R (1987) The hydrogeology, hydrochemistry and environmental isotopes of the Campaspe River aquifer system, north-central Victoria, Australia. J Hydrol 95: 63–86

Blackburn G, McLeod S (1983) Salinity of atmospheric precipitation in the Murray Darling Drainage Division, Australia. Austr J Soil Res 21: 400–434

Brown CM (1989) Structural and stratigraphic framework of groundwater occurrence and surface discharge in the Murray Basin, southeastern Australia. Bur Min Res J Austr Geol Geophys 11: 127–146

Bureau of Meteorology (2005) Commonwealth of Australia Bureau of Meteorology. http://www.bom.gov.au/

Calf GE, Ife D, Tickell S, Smith LW (1986) Hydrogeology and isotope hydrology of Upper Tertiary and Quaternary aquifers in northern Victoria. Austr J Earth Sci 33: 19–26

Cartwright I, Weaver TR (2005) Hydrogeochemistry of the Goulburn Valley region of the Murray Basin, Australia: implications for flow paths and resource vulnerability. Hydrogeol J 13: 752–770

Cartwright I, Weaver TR, Fulton S, Nichol C, Reid M, Cheng X (2004) Hydrogeochemical and isotopic constraints on the origins of dryland salinity, Murray Basin, Victoria, Australia. Appl Geochem 19: 1233–1254

Cartwright I, Weaver TR, Fifield LK (2006) Cl/Br Ratios and Environmental Isotopes as Indicators of Recharge Variability and Groundwater Flow: An Example from the Southeast Murray Basin, Australia. Chemical Geology 231: 38–56

Clark ID, Fritz P (1997) Environmental Isotopes in Hydrogeology. Lewis, New York, USA. 328p

Coplen, TB (1988) Normalization of oxygen and hydrogen isotope data. Chem Geol 72: 293–297

Davis SN, Cecil LD, Zreda M, Moysey S (2001) Chlorine-36, bromide, and the origin of spring water. Chem Geol 179, 3–16

Davis SN, Whittemore DO, Fabryka-Martin J (1998) Uses of chloride/bromide in studies of potable water. Ground Water 36, 338–351

Dimos A, Chaplin H, Potts I, Reid M, Barnwell K (1994) Bendigo Hydrogeological Map (1:2,50,000 scale). Austr Geol Surv Org, Canberra, Australia.

Dogramaci SS, Herczeg AL (2002) Strontium and carbon isotope constraints on carbonate-solution interactions and inter-aquifer mixing in groundwaters of the semi-arid Murray Basin, Australia. J Hydrol 262: 50–67

Drever JI (1997) The geochemistry of natural waters: surface and groundwater environments. Prentice-Hall, New Jersey, USA. 436p

Edmunds WM, Bath AH, Miles DL (1982) Hydrochemical evolution of the East Midlands Triassic sandstone aquifer, England. Geochim Cosmochim Acta 46: 2069–2082

Evans WR (1988) Shallow groundwater salinity map of the Murray Basin (1:10,00,000 scale). Bur Min Res, Canberra, Australia

Evans WR, Kellett JR (1989) The hydrogeology of the Murray Basin, southeastern Australia. Bur Min Res J Austr Geol Geophys 11: 147–166

Fabryka-Martin J, Whittemore DO, Davis SN, Kubik PW, Sharma P (1991) Geochemistry of halogens in the Milk River aquifer, Alberta, Canada. Appl Geochem 6: 447–464

Ghassemi F, Jakeman AJ, Nix HA (1995) Salinisation of land and water resources: Human causes, extent, management, and case studies. Univ. New South Wales Press, Sydney, Australia, 526 p

Goller R., Wilcke W, Leng MJ, Tobschall HJ, Wagner K, Valarezo C, Zech W (2005) Tracing water paths through small catchments under a tropical montane rain forest in south Ecuador by an oxygen isotope approach. J Hydrol 308: 67–80

Gonfiantini R (1986) Environmental isotopes in lake studies. In: Fritz P, Fontes JC (Eds) Handbook of Environmental Isotope Geochemistry, vol 2, the Terrestrial Environment. Elsevier, Amsterdam, Netherlands. pp. 113–168

Hannam K, Cartwright I, Weaver TR (2004) Hydrogeochemistry of the Lake Cooper region, Murray Basin. In: Wanty RB, Seal RR (Eds) Proceedings of Water-Rock Interaction 11. Balkema, Netherlands. pp. 405–410

Harrington GA, Herczeg AL (2003) The importance of silicate weathering of a sedimentary aquifer in arid Central Australia indicated by very high $^{87}Sr/^{86}Sr$ ratios. Chem Geol 199: 281–292

Hendry MJ, Schwartz FW, Robertson C (1991) Hydrogeology and hydrochemistry of the Milk River aquifer system, Alberta, Canada: a review. Appl Geochem 6, 369–380

Hennessy J, Reid M, Chaplin H (1994) Wangaratta Vic/NSW Hydrogeological Map of the Murray Basin (1:250,000 scale). Geoscience Australia, Canberra, Australia

Herczeg A, Edmunds WM (2000) Inorganic ions as tracers. In: Cook P, Herczeg A (Eds) Environmental Tracers in Subsurface Hydrology. Kluwer Academic Publishers, Boston. pp. 31–77

Herczeg AL, Barnes CJ, Macumber, PG, Olley JM (1992) A stable isotope investigation of groundwater-surface water interactions at Lake Tyrrell, Victoria. Chem Geol 96: 19–32

Herczeg AL, Dogramaci SS, Leaney FW (2001) Origin of dissolved salts in a large, semi-arid groundwater system: Murray Basin, Australia. Marine Freshwater Res 52: 41–52

Herczeg AL, Simpson HJ, Mazor E (1993) Transport of soluble salts in a large semiarid basin: River Murray, Australia. J Hydrol: 144, 59–84

Herczeg AL, Torgersen T, Chivas AR, Habermehl MA (1991) Geochemistry of ground waters from the Great Artesian Basin, Australia. J Hydrol: 126, 225–245.

Ivkovic KM, Watkins KL, Cresswell RG, Bauld J (1998) A Groundwater Quality Assessment of the Upper Shepparton Formation Aquifers: Cobram Region, Victoria. Austr Geol Surv Org. Record 1998/16. Canberra, Australia. 89p

Kloppmann W, Négrel Ph, Casanova J, Klinge H, Schelkes K, Guerrot C (2001) Halite dissolution derived brines in the vicinity of a Permian salt dome (N German Basin). Evidence from boron, strontium, oxygen, and hydrogen isotopes. Geochim Cosmochim Acta 65: 4087–4101

Lawrence CR (1988) Murray Basin. In: Douglas JG, Ferguson JA (Eds). Geology of Victoria. Geol Soc Austr (Victoria Div), Melbourne, Australia. pp. 352–363

Leaney FW, Herczeg AL (1999) The origin of fresh ground water in the southwest Murray Basin and its potential for salinization. CSIRO Land and Water Technical Rep 7/9. www.clw.csiro.au/publications/technical99/tr7-99.pdf

Leaney FW, Herczeg AL, Walker GR (2003) Salinization of a fresh palaeo-ground water resource by enhnaced recharge. Ground Water 41: 84–92

Love AJ, Herczeg AL, Armstrong D, Stadter F, Mazor E (1993) Groundwater flow regime within the Gambier Embayment of the Otway Basin, Australia: evidence from hydraulics and hydrochemistry. J Hydrol 143: 297–338

Macumber PG (1991) Interaction between groundwater and surface water systems in northern Victoria. Victoria Dept Cons Env Melbourne, Australia. 345p

McCaffrey MA, Lazar B, Holland HD (1987) The evaporation path of seawater and the coprecipitation of Br^- and K^+ with halite. J Sed Petrol 57: 928–937

McGuire KJ, DeWalle DR, Gburek WJ (2002) Evaluation of mean residence time in subsurface waters using oxygen-18 fluctuations during drought conditions in the mid-Appalachians. J Hydrol 261: 132–149

Parkhurst DL, Appelo CAJ (1999) User's guide to PHREEQC (v.2) – a computer program for speciation, batch-reaction, one-dimensional transport, and inverse geochemical calculations: US Geol Surv Water-Resources Investigations Report 99-4259. Washington, DC, USA. 312p

Radke BM, Ferguson J, Cresswell RG, Ransley TR, Habermehl MA (2000) Hydrochemistry and implied hydrodynamics of the Cadna-owie – Hooray Aquifer, Great Artesian Basin, Australia. Bureau of Rural Sciences, Canberra. 229p

Ragab R, Cooper JD (1993) Variability of unsaturated zone water transport parameters: implications for hydrological modelling. 1. In situ measurements. J Hydrol 148: 109–131

Stephenson AE, Brown CM (1989) The ancient Murray River system. Bur Min Res J Austr Geol Geophys 11: 387–395

Tickell SJ (1978) Geology and hydrogeology of the eastern part of the riverine plain in Victoria. Geol Surv Victoria Report 1977–8. Melbourne, Australia. 73p

Tickell SJ (1991) Shepparton geological report. Geol Surv Victoria Report 88. Melbourne, Australia. 75p

Tickell SJ, Humphries J (1985) Hydrogeological map of Bendigo and part of Deniliquin. Dept Ind Tech Res Victoria. Melbourne, Australia

Tickell SJ, Humphries J (1986) Groundwater resources and associated salinity problems of the Victoria part of the Riverine Plain. Geol Surv Victoria Report 84. Melbourne, Australia. 104p

Weaver TR, Bahr JM (1991) Geochemical evolution in the Cambrian-Ordovician sandstone aquifer, eastern Wisconsin; 2, Correlation between flow paths and ground-water chemistry. Ground Water 29: 510–515

Weaver TR, Frape SK, Cherry JA (1995) Recent cross-formational fluid flow and mixing in the shallow Michigan Basin. Geol Soc Am Bull 107: 697–707

Vogel JC (1970) Groningen radiocarbon dates IX. Radiocarbon, 12, 444–471

CHAPTER 5

The South African groundwater decision tool

Ingrid Dennis* and Sonia Veltman**
*International Association of Hydrogeologists, Institute for Groundwater Studies,
University of the Free State, Bloemfontein, South Africa
**Department of Water Affairs and Forestry, Pretoria, South Africa

ABSTRACT: Sustainability, equity and efficiency are identified as central guiding principles in the protection, use, development, conservation, management and control of water resources in South Africa. These principles recognise:

- the basic human needs of present and future generations,
- the need to protect water resources,
- the need to protect aquatic ecosystems,
- the need to share some water resources with other countries,
- the need to promote social and economic development through the use of water, and
- the need to establish suitable institutions in order to achieve the above-mentioned principles

In order to implement these principles, the South African government needs to ensure that the tools and expertise required are available. The South African Groundwater Decision Tool (SAGDT) is a risk-based methodology developed to assist regional and local water resource managers in decision making with regard to aquifer use, protection and management.

This paper introduces the SAGDT, which utilises fuzzy logic rules to determine the sustainability of groundwater resource, risk of contamination, human health risks associated with contaminated groundwater and the impacts (quantity and quality) of groundwater on aquatic ecosystems. In all the aspects of the SAGDT groundwater flow systems play a role and therefore these are taken into account together with the fuzzy logic rules.

The SAGDT was applied in the Kromme River, results indicates that there is a 39% risk of failure of boreholes BH3 and BH4 over a 2 year period as the groundwater flow system is not being able to sustain present borehole extraction rate. This risk is increased to 60% for boreholes BH1 and BH2 as they are only 20 m apart. If groundwater extraction due to Black Wattle trees is included in the simulation in the vicinity of BH3, an increased risk of failure of 52% is added. This suggests that clearing Black Wattle trees may reduce the risk of borehole failure and restore natural groundwater flow patterns.

Key words: Risk assessment, groundwater management, groundwater protection, South Africa.

1 INTRODUCTION

Water of acceptable quality is both necessary for the improvement of the quality of life and essential to the maintenance of all forms of life. A balance has to occur between the protection, use, development, conservation, management and control of water resources.

Understanding groundwater flow systems is key to determining this balance. According to South African legislation, the following aspects need to be taken into account:

- The basic human needs of present and future generations
- The need to protect water resources
- The need to protect aquatic ecosystems
- The need to share some water resources with other countries
- The need to promote social and economic development through the use of water, and
- The need to establish suitable institutions in order to achieve the above-mentioned aspects.

The South African Groundwater Decision Tool (SAGDT) is designed to provide methods/tools to assist groundwater professionals and regulators in making informed decisions concerning groundwater use, management and protection, while taking into account that groundwater forms part of an integrated water resource. The SAGDT is spatially-based software, which includes:

- A geographic information system (GIS) interface allows a user to import shape files, various computer aided design (CAD) formats and geo-referenced images. The GIS interface also provides for spatial queries to assist in the decision-making process. The GIS interface contains a default set of shape files depicting various hydrogeological parameters across South Africa
- Risk assessment interface. The SAGDT introduces fuzzy logic based risk assessments to assist in decision making by systematically considering all possibilities. Included risk assessments relates to the sustainability of a groundwater resource, contamination of a groundwater resource, human health risks associated with a contaminated groundwater resource and impacts of changes in groundwater (quantity and/or quality) on aquatic ecosystems
- Third-party software includes a shape file editor, an interpolator and a groundwater dictionary, which includes a definition, a description on why the term is important when considering groundwater. Graphics are used to assist in understanding the terminology
- A report generator, which automatically generates documentation concerning the results of the risk assessment performed and the input values for the risk assessment, and
- A scenario wizard is available for the novice to obtain step by step instructions in setting up a scenario.

The SAGDT allows problem solving at a regional scale or a local scale, depending on the problem at hand, and as such groundwater flow systems must be understood at various scales. This paper focuses on the SAGDT, and more specifically on the risk assessment methodologies applied in the SAGDT. A case study is used to demonstrate the functioning of the SAGDT and the importance of the groundwater flow systems when managing groundwater resources.

2 METHODOLOGY

2.1 *Preamble*

According to the specification provided by the South African Department of Water Affairs and Forestry, the developed software has to adhere to the following:

- Be made of a standard system of consistent methods/rules to guide planning and decision making about water resources

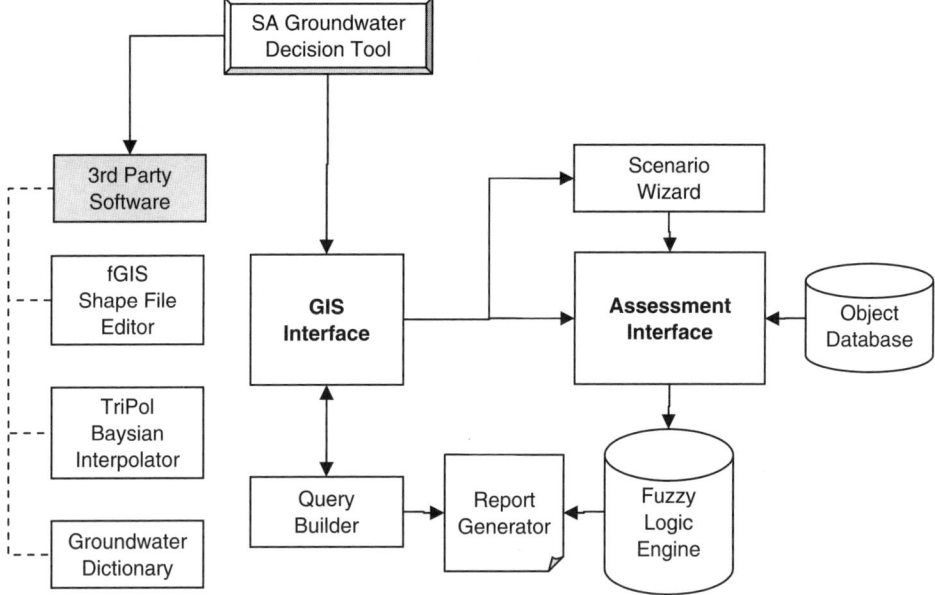

Figure 1. High level architecture of the SAGDT.

- Allow transparency, accountability and long-term goal setting to be incorporated into water resource management, and
- Calculate the level of confidence of results obtained.

2.2 *System architecture*

The high-level architecture of the SAGDT is shown in figure 1, and the sub-systems will be discussed in more detail in the sections to follow.

2.3 *GIS and Assessment Interface*

The SAGDT comprises the following main components:

- GIS Interface – Most national GIS based groundwater datasets have been included in the SAGDT and can be accessed in a GIS environment. This provides the user with values for essential parameters on a regional scale. Through the selection of a point, a GIS object is created for the assessment, with all the required parameters extracted from the GIS environment. It continues without indicating that low-confidence values are assigned to the GIS parameters, since hydrogeological parameters can change over a short distance. The aim of the GIS object is to provide the user with an estimate for parameters for which no data are available.
- Assessment Interface – The assessment interface is a CAD environment, supported by a finite difference flow and transport model. A scenario is built through the use of a library of objects presented in table 1. An object represents a physical entity for example

Table 1. Library of objects.

Name	Object	Input	Calculation	Output
Opencast Mine	Area	Recharge (R), Area of mine, Inflow of groundwater into the mine (C), Outflow of groundwater out of the mine dose (O), Volume of mine, Storativity of spoils, Sulphate generation and Low flow in river	Decant rate = (R × area of mine) + I − O $$\text{Time} = \frac{\text{volume of open cast mine} \times \text{storavity of spoils}}{I + R}$$ Load of sulphate at river = Concentration × decant rate $$\text{Mixing} = \frac{\text{Load of sulphate at river}}{\text{Low flow in river}}$$	Decant Parameters
Dam	Area	Closed polygon points, Vegetation type, Perennial state and Root depth	No calculations performed as this feeds directly into ecological risk assessment (fuzzy logic) calculations and the finite difference model	Input Parameters
River	Area	Closed polygon points, Vegetation type, Perennial state and Root depth	No calculations performed as this feeds directly into ecological risk assessment (fuzzy logic) calculations and the finite difference model	Input Parameters
Flow Boundary	Area	Closed polygon points, Transmissivity and Storativity	No calculations performed as this feeds directly into the finite difference model	None
Wetland	Area	Closed polygon points, Vegetation type, Perennial state and Root depth	No calculations performed as this feeds directly into ecological risk assessment (fuzzy logic) calculations and the finite difference model	Input Parameters
Borehole	Point	Coordinate, Name, Water strike, Extraction rate, Blow yield and Recharge	No calculations performed as this feeds directly into the sustainable risk assessment (fuzzy logic) calculations and finite difference model	Input Parameters
Toxin	Point	Total dose (Dose), Max concentration (C), Average intake rate (IR), Exposure duration (ED), Average daily dose (ADD), Average body weight over exposure period (BW) and Reference dose (RfD)	Dose = C × IR × ED $$\text{ADD} = \frac{\text{Dose}}{\text{BW} \times \text{ED}}$$ $$\text{Risk} = \frac{\text{ADD}}{\text{RfD}}$$	Risk of health impacts on humans due to toxins

		Input Parameters		
Carcinogen	Point	Total dose (Dose), Max concentration (C), Average intake rate (IR), Exposure duration (ED), Average daily dose (ADD), Average body weight over exposure period (BW), Lifetime average daily dose (LADD) and Cancer potency factor (CPF)	$\text{Dose} = C \times IR \times ED$ $ADD = \dfrac{\text{Dose}}{BW \times ED}$ $LADD = \dfrac{\text{Total dose}}{BW \times \text{lifetime}}$ $\text{Risk} = 1 - e^{-LADD \times CPF} \approx LADD \times CPF$	Risk of humans developing cancer
Population	Point	Population size, Percentage of population under the age of 2 years, Percentage of the population over the age of 60 years and Percentage of people dependant on groundwater	No calculations performed as this feeds directly into the risk assessment (fuzzy logic) calculations	
Radiation	Function	Total dose (dose), Max concentration (C), Risk coefficient (r) and Exposure duration (ED)	$\text{Risk} = r \times \text{Dose}$ For inhalation and ingestion $\text{Risk} = r \times C \times ED$ For submersion risk	Risk of radiation impacting on human health
Microbial	Function	Number of organisms (N) and Parameters characterised by dose-response curves (α, β and r)	The single-hit exponential model $\text{Risk} = 1 - e^{(-rN)}$ OR the beta-distributed model $\text{Risk} = 1 - \left[1 + \left(\dfrac{N}{\beta}\right)\right]^{-\alpha}$	Risk or probability of infection
Chloride	Function	Mean annual precipitation (MAP), Chloride in rainfall (Rcl) and Chloride in groundwater (Gcl)	$\text{Recharge} = \left(\dfrac{Rcl \times MAP}{Gcl}\right) \times 100$	Percentage groundwater recharge
Earth	Function	Rainfall data, Water level data (h), Specific yield (S), Change in water level head with time (dh/dt) and Drainage resistance (DR)	$S \dfrac{dh}{dt} = \text{Recharge} - \left(\dfrac{h}{DR}\right)$	Groundwater recharge

(Continued)

Table 1. (Continued)

Name	Object	Input	Calculation	Output
Herold	Function	Total flow during month i (Qi), Groundwater contribution (QGi), Surface runoff (QSi), Minimum groundwater flow (GMAX), Groundwater decay factor ($0 < $ DECAY < 1) and Groundwater growth factor ($0 < $ PG > 1)	$Qi = Qgi + Qsi$ $QSi = Qi - QGMAX$ (for $Qi > QGMAX$) or $QSi = 0$ (for $Qi \leq QGMAX$) and hence $QGi = Qi - Qsi$ $GGMAX_i = DECAY.GGMAY_{i-1} + PG.QS_{i-1}/100$	Groundwater contribution to base-flow
Logan	Function	Discharge rate (Q) and Drawdown (s)	$T \approx 1.22 \dfrac{Q}{s}$	Transmissivity
Slug	Function	Recession time (t)	$Q = 117155.08 t^{-0.824}$ $T = 10 \times Q$	Transmissivity
Cooper-Jacob	Function	Discharge rate (Q), Drawdown (s), Time since start of pumping test (t), Storativity (S) and Radius of borehole (r)	$s = \dfrac{2.3Q}{4\pi T} \log \dfrac{2.25Tt}{r^2 S}$	Transmissivity
Reserve	Function	Number of people dependent on groundwater, Recharge and Groundwater contribution to base-flow	$\text{Reserve} = \left(\dfrac{\text{No of people} \times \text{Basic human needs} + \text{Base-flow}}{\text{Recharge}} \right) \Big/ 100$	Groundwater reserve expressed as a percentage of recharge

a borehole, contaminant source or wetland. Objects can also be a tool to analyse data such as the Cooper-Jacob method. Each object type has a specific confidence assigned to it, depending on the method applied by the object and the level of data required to perform the calculation. As an example, consider a slug test and a pumping test analysis to determine a transmissivity value, where the pumping test data would yield a higher level of confidence than the other method. Each object placed in the assessment interface is used to refine the data provided by the GIS object. Each scenario is represented by an object tree which is analysed by the SAGDT to determine the various risk categories involved. The GIS object is the parent of all object trees and an object tree allows inheritance. Inheritance is used when a certain parameter value is not available at a lower level in the tree, thus inheriting that parameter value from a higher level in the tree. This explains why the GIS object with default values for all parameters in the SAGDT environment is used as a parent for each scenario. As mentioned earlier the user should make use of the object library to refine the parameters presented in the GIS object as far as possible to yield accurate assessment results.

2.4 *Object library*

The objects used to create a scenario are stored in the object library. An object must adhere to a generic framework, allowing the addition of objects to the library without any changes to the source code of the SAGDT. The object library also contains all the associated help documentation explaining the application and methodology of each object. A summary of all the objects in the current version of the SAGDT are presented in table 1.

The objects are divided into three main categories:

- Area objects such as a river or wetland
- Point objects such as a borehole or population
- Function objects, which are usually methods such as Cooper-Jacob analysis of pumping test data to determine aquifer transmissivity.

The different object types will allow for the assessment interface to have all or combinations of following properties, depending on the object selections comprising the scenario:

2.4.1 *Sustainability category*

There are many definitions for groundwater sustainability. Sharp (1998), for example, defined the sustainable yield of groundwater as the minimisation of potentially negative effects on an aquifer, so that it can be utilised at an acceptable range of levels for a very long period of time. Merrick (2000) stated that sustainable yield is that proportion of the long-term annual recharge that can be extracted each year without causing unacceptable impacts on groundwater users or other components of the environment (such as aquatic ecosystems). Van Tonder (2001) builds on this by adding aspects such as time, position of the pump or the main water strike and borehole construction and management. A groundwater quantity or sustainability risk assessment has therefore been designed to determine the risks of failure when extracting water from an aquifer. The factors taken into consideration in this risk assessment are:

- Recharge, which is an important factor according to the definitions of sustainable yield.
- Water strike/depth of main fracture, which determines the amount of drawdown possible in a borehole. According to Van Tonder's (2001) definition of sustainable yield, it is

important not to extract a quantity of water such that the water level reaches the water strike or pump. In the case of a porous aquifer a certain percentage of the saturated thickness above the pump would be an indication of the possible drawdown.
- The drawdown in the borehole under investigation must not reach the main water strike or pump. This drawdown is calculated taking into account the groundwater flow systems and the influence of other extraction boreholes. Aquifer tests are used to obtain information concerning aquifer parameters in order to calculate drawdown.
- The period for which the users wish to extract is important. Calculations show that, the longer the period of extraction, the larger the impact on the groundwater flow system.

It is important to note that the sustainability category only considers the quantity of water available, however if there are probable contaminants present, a contaminant assessment must be considered together with a sustainability assessment to determine the final risk of system failure.

2.4.2 *Contamination category*
Groundwater contamination can be defined as the introduction of any substance into a groundwater system through human action. The following information is taken into account in the contamination assessment:

- The contaminant and associated guidelines (such as drinking water guidelines) as these dictate the fuzzy logic rules.
- Duration of contamination: if the contamination results from a single (once-off) spill, the impact will probably be smaller than that resulting from continuous contamination.
- Factors that influence the movement of a contaminant such as the groundwater flow system, matrix diffusion and dispersion coefficient.

2.4.3 *Groundwater vulnerability (to contamination) category*
Groundwater vulnerability represents the intrinsic characteristics that determine the aquifer's sensitivity to the adverse effects resulting from the imposed contaminant (Lynch et al., 1994).

The parameters needed for describing groundwater vulnerability to contamination are:

- Depth to groundwater and character of the unsaturated zone: this gives an indication of the distance and time required for the contaminant to move through the unsaturated zone to the saturation level, taking into account the character of the unsaturated zone (for example if it has a porous or fractured nature or if it is unconfined, semi-confined or confined). Soil media that can form the upper portion of the unsaturated zone must also be taken into account at this point. The various physical and chemical properties of soils can either retard or accelerate the movement of a contaminant. Typical South African soil types have been classified according to these properties and this classification system is used in the SAGDT.
- Recharge: the primary source of groundwater is precipitation which aids in the movement of a contaminant into the aquifer.
- Aquifer media: the consolidated or unconsolidated rock matrices that serve as water-bearing units. In this approach, the fractures that occur in the rock matrix can also be taken into account.

- Topography: will give an indication on whether a contaminant might runoff or remain on the surface long enough to infiltrate into the groundwater. This is based on the assumption the steeper the slope, the more runoff and less infiltration into the groundwater system.

Lynch et al. (1994) classified numerous South African aquifers according to the above-mentioned parameters. This is in turn converted the fuzzy logic rules within the SAGDT.

2.4.4 Health category

A groundwater health risk assessment can be defined as a qualitative or quantitative process to characterise the probability of adverse health effects associated with measured or predicted levels of hazardous agents in groundwater (Dennis et al., 2002). Once a contaminant is released into the groundwater, its resultant concentration to reach the human body is dependent upon the physical and chemical properties of both the contaminant and the groundwater.

In addition, the concentrations found in a human are subject to the person's exposure to groundwater, it also depends on the diet, age, health status, activities, among other causes. Exposure is defined by the frequency, magnitude and duration of contact with the contaminant (Schwab and Genthe, 1998). Frequency refers to whether a person is exposed daily or just occasionally. The magnitude refers to the amount of exposure. The duration refers to whether any single exposure episode may last for minutes, hours, days or years. Once the contaminant is inside the body, it may be further transformed via metabolism or detoxification. Children, the elderly and those with chronic conditions, for instance, react differently to the same dose than the average, healthy middle-aged adult (Schwab and Genthe, 1998). The impact of contaminants for the various scenarios are characterised in a health risk assessment. The following aspects are therefore taken into account when performing the health risk assessment:

- Toxicity of the contaminant: When exposed to toxic chemicals, there are numerous health effects that vary from mild headaches to death, all of which need to be taken into account in a risk assessment.
- Carcinogeneity of a contaminant: Exposure to certain chemicals can cause some forms of cancer, and therefore the carcinogeneity of a chemical needs to be taken into account when conducting a health risk assessment.
- Possibility of infection: Allows the user to obtain an idea of the risks involved in human exposure to a variety of bacteria, viruses and protozoa.
- Radiation exposure can result in delayed effects such as cancer.
- Exposure to a contaminant: This establishes whether exposure to a chemical or microbiological agent can cause harm. To determine exposure, it is necessary to combine an estimation of groundwater concentrations of the hazards with demographic or behavioural descriptions of the exposed population.
- Population exposed to a contaminant: The population is composed of groups who differ in their vulnerability to health hazards. Babies are for example more susceptible to infection because of their lack of immunity.

The above-mentioned aspects are taken into account in standard risk assessment methodologies such as those specified by the United States Environmental Protection Agency

(Environmental Protection Agency, 1989). It is important to note that no fuzzy logic is included in the health risk assessments.

2.4.5 *Ecological category*

The National Water Act (1998) of South Africa is based on a number of principles, one of which is that the quantity, quality and reliability of water required to maintain the ecological functions of aquatic and associated ecosystems. The Water Act focuses on aquatic and associated ecosystems and therefore only these have been included in the risk assessment process.

Ecological risk assessments differ from health risk assessments in several significant ways. For ecosystems, the risk assessment methodology must consider effects beyond just individual organisms or a single species. With ecosystems, some sites and types are more valuable and vulnerable than others. Accommodating these factors complicates ecological risk assessments and renders them more subjective. Unfortunately, there are limited data available concerning South African aquatic ecosystems. The SAGDT therefore only consider factors such as:

- Ecological importance and sensitivity: Ecological importance of a river is an expression of its importance to the maintenance of ecological diversity and functioning on local and wider scales. Ecological sensitivity (or fragility) refers to the system's ability to resist disturbance and its capability to recover from disturbance once it has occurred (resilience) (Resh *et al.*, 1988; Milner 1994).
- Dependency of vegetation on groundwater: The degree of dependency of vegetation on groundwater as a source of water and survival is taken into account in the fuzzy logic rules. The dependency level ranges from vegetation entirely dependent on groundwater systems to those which do not use groundwater at all.
- Groundwater-surface water interaction: the link between the groundwater and surface water systems must be established as one of the indicators of the role groundwater plays in the sustainability of the ecosystem.
- Groundwater extraction versus groundwater contribution to base-flow: the impacts of groundwater extraction must be compared to the volume of groundwater flowing into the system.
- Aquatic ecosystem guidelines: these guidelines provide an indication as to when the groundwater quality becomes unacceptable for the ecosystems present.

2.5 *Fuzzy logic risk assessment engine*

Conventional set theory states that an element is either a member of a set or not. Fuzzy logic is an extension of conventional set theory, enabling an element to belong to a set to a degree. The degree of membership is a function that defines the membership of an element to a set according to the value of the element, as shown in figure 2. Membership is expressed as a value between 0 and 1. Zero implies 0% membership and 1 implies 100% membership. Note that, in most cases, the membership functions of the two sets will be inverses. The membership function is selected by an expert in the field of study. Linear membership functions are seldom used in practice, in contrast with sinusoidal functions, which are very popular. In most cases, risk analysis will involve more than one input to be considered.

Degree of membership

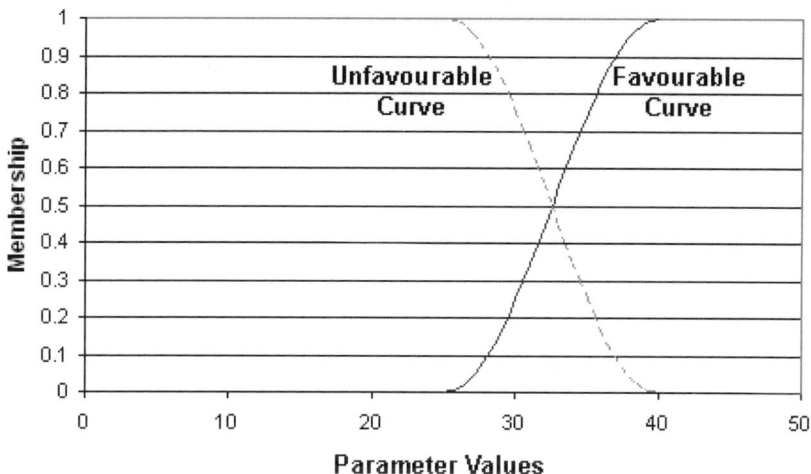

Figure 2. Fuzzy logic membership example.

Table 2. Rule set.

Rule No	Weight	Input 1	Input 2	Input 3
1	0.0	Favourable	Favourable	Favourable
2	?	Favourable	Favourable	Unfavourable
3	?	Favourable	Unfavourable	Favourable
4	?	Favourable	Unfavourable	Unfavourable
5	?	Unfavourable	Favourable	Favourable
6	?	Unfavourable	Favourable	Unfavourable
7	?	Unfavourable	Unfavourable	Favourable
8	1.0	Unfavourable	Unfavourable	Unfavourable

Fuzzy logic makes it possible to generate a set of decision rules, according to the number of inputs. These rules must then be evaluated by an expert in the field of study. The number of rules generated is given by Equation 1.

$$n = 2^{inputs} \tag{1}$$

where n represents the number of rules generated. The rules consist of all possible binary combinations of the respective inputs, with a weight assigned to each rule representing the risk. Consider the rule set presented in table 2 for three input parameters. All rules are read in the same fashion, and an expert must evaluate each rule individually to assign the appropriate risk. Rule 3 of the rule set in table 2 will read as follows:

If Input 1 is favourable, Input 2 is unfavourable and Input 3 is favourable what would be the risk where 1 represents 100%?

For each input, a membership function must be defined, with a favourable and unfavourable limit defining the two sets. One function will represent the favourable set and the other the unfavourable set. Thus, for each input, a favourable and an unfavourable value can be read from the membership functions. For each input, the table of decision rules is then populated with the respective favourable and unfavourable degree of membership, and the scenario risk calculated using Equation 2.

$$\%\text{Risk} = \frac{\sum_1^n Wn^* \min(DOM)}{\sum_1^n \min(DOM)} \times 100 \tag{2}$$

Where
n = number of rules
DOM = degree of membership
Wn = weight of rule n

Note that the minimum function must return the minimum value of all inputs for each rule. The fuzzy logic rule sets and the membership functions for each parameter are stored in a database and are only available for editing to users with administrative rights to the application.

2.6 Risk profile report

After a risk analysis is done, the SAGDT will produce a profile report containing the following information:

- Map of area
- Summary of object properties and calculated values
- Risk assessment per applicable category

2.7 Scenario wizard

The scenario wizard consists of a few typical scenarios that assist the novice user through a step by step approach in setting up these scenarios. Examples of the wizard scenarios include the following, to name a few:

- Determine a sustainable extraction rate from a borehole
- Contamination close to a borehole used for human consumption
- Decommissioning of an opencast mine

3 CASE STUDY

3.1 Preamble

The Kromme River Catchment is located in the Eastern Cape, South Africa, to the west of Jeffreys Bay. The Kromme River is located in a narrow plane between the Suuranys and Tsitsikamma mountains, is approximately 95 km long and drains a catchment area of 1,125 km². It runs in an easterly direction and exits into the Indian Ocean at St Francis Bay.

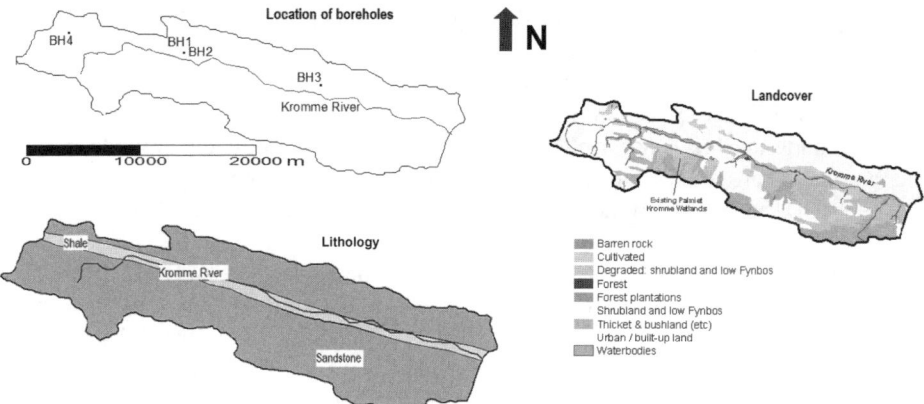

Figure 3. Main lithological features, borehole location and land cover of the kromme river catchment.

The groundwater flows towards the Kromme River. The study area focuses along the first 30 km of the Kromme River (figure 3). There are approximately 259 people living in this area, some of which are dependent on groundwater for their basic human needs. In addition, there are sensitive groundwater-fed wetlands along this stretch of the Kromme River. Bulrushes (*Prionium serratum*) have various important functions in the wetland systems of the Kromme River, such as maintaining surface water flow, reducing erosion and maintaining ecological systems. The area is also home to indigenous fynbos (such as *Protea caffra caffra* and *Protea repens*), referring to a distinctive community of plants found in the South Western Cape. Many of these plants have small, fine stems and leaves. The vegetation has a bushy appearance.

Agricultural activities are destroying the riparian zone and alien vegetation (especially Black Wattle trees – *Acacia mearnsii*) and impacting groundwater and wetlands, and therefore the Kromme River flow. As the Kromme River is one of the main sources of clean drinking water for growing urban areas in adjacent catchments, it is vital that these groundwater-fed wetlands be preserved as they in turn preserve ecosystems related to the river flow.

The aim of this case study is to determine the risks associated with the impacts of farming activities and alien vegetation on the groundwater system and associated wetlands.

3.2 Sustainable risk assessment

The sustainability of four basic human needed boreholes is tested using the SAGDT. The yields of the boreholes (≈ 100 m deep) vary between 0.4 l/s to 0.8 l/s. The Table Mountain Sandstone Group (TMG) covers the bulk of the study area (figure 3) and consists almost entirely of metaquartzites, derived from metamorphism of medium-grained arenaceous sandstones. The contact between the TMG and the Bokkeveld shale is weathered and fractured. Large-scale regional fault zones play a major role in the hydrogeology of the area creating complex groundwater flow patterns. Most aquifer units associated with the TMG are considered to have a semi-confined hydraulic response. The average annual precipitation is 774 mm. The recharge for the study area is calculated to be 3.5% of the mean annual precipitation. Chemical analyses, groundwater levels and groundwater gradients indicate that the groundwater is recharging the Kromme River. Topographic divides are assumed as no-flow boundaries. It is important to note that this assumption is based on regional groundwater levels. However the position of this boundary can change with time, with for example groundwater extraction. The depth to groundwater ranges from 3–12 m. The shallower groundwater levels occur close to the river, indicating a possible groundwater discharge zone. There is uncertainty concerning the water strike depth.

The SAGDT indicates that there is a 39% risk that boreholes BH3 and BH4 will fail over a period of 2 years due to the groundwater flow system not being able to sustain the borehole's extraction rates. This risk is increased to 60% for boreholes BH1 and BH2 due to these boreholes only being 20 m apart. If groundwater extraction due to Black Wattle trees is included in the simulation in the vicinity of BH3, results indicate an increased risk of failure of 52% due to the presence of Black Wattle trees. Therefore it can be concluded that clearing Black Wattle trees in the area can reduce the risk of borehole failure and restore natural groundwater flow patterns.

3.3 *Ecological risk assessment*

As the South African National Water Act (1998) focuses on aquatic ecosystems, the SAGDT also only focuses on these ecosystems and more specifically on the vegetation in the wetlands and riparian zones. In the case of the Kromme River only the impacts of the reduced flow towards these systems are considered. There is a 70% risk of failure of these ecosystems due to the impacts of alien vegetation and groundwater extraction.

4 DISCUSSION AND CONCLUSIONS

The SAGDT concept can be a powerful groundwater management tool. However it is important to note that the underlying foundation of this tool is the understanding of groundwater flow systems and capturing the knowledge of experts who understand the functioning of these systems. A risk based fuzzy logic assessment interface is built on this foundation, thereby providing a common framework for all groundwater practitioners in South Africa, in which they can perform groundwater risk assessments that relate to policy.

The SAGDT can calculate risks taking the following into account:

- sustainability of a borehole, borehole-field or groundwater flow system
- groundwater vulnerability to contamination
- contamination of a borehole, borehole-field or groundwater flow system
- impacts of changes in a groundwater flow system on the aquatic ecosystems.

The SAGDT also acts as a groundwater educational environment, due to the extensive groundwater dictionary and object help files available.

ACKNOWLEDGMENTS

A special thanks to the South African Water Research Commission and the Department of Water Affairs and Forestry for financially supporting the development of the SAGDT.

REFERENCES

Dennis I, Van Tonder GJ and Riemann K (2002) Risk based decision tool for managing and protecting groundwater resources. Water Research Commission Report 969/1/02. ISBN No. 1-86845-924-1, Pretoria, South Africa.

Environmental Protection Agency (1989) Risk assessment guidance for Superfund, vol 1. Human health manual (part A), Rep EPA/540/1-89/002. Office of Emergency and Remedial Response, Washington, D.C.

Lynch SD, Reynders AG and Schulze RE (1994) Preparing input data for a national-scale groundwater vulnerability map of Southern Africa. Water SA vol 20, no 3, p 239–246.

Merrick NP (2000) Assessing uncertainty in sustainable yield. In Sililo et al. (eds) GROUNDWATER: Past Achievements and Future Challenges. Proc of the XXX IAH Congress, South Africa. p 401–406.

Milner AM (1994) System recovery. In: P Calow & GE Petts (eds.). The rivers handbook. Vol. 2. Blackwell Scientific Publications. London.

National Water Act. Act 36 of 1998, Pretoria, South Africa.

Resh VH, Brown AV, Covich AP, Gurtz ME, Li HW, Minshall GW, Reice SR, Sheldon AL, Wallace JB and Wissmar RC (1988) The role of disturbance theory in stream ecology. Journal of the North American Benthological Society, vol 7 p 433–455.

Schwab M and Genthe B (1998) Environmental Health Risk Assessment: A primer for South Africa. CSIR report no. ENV/S-I 98029, Environmentek, CSIR, PO Box 320, Stellenbosch, South Africa p 1–14.

Sharp JM (1998) Sustainable Groundwater Supplies – an evolving issue: Examples from Major Carbonate Aquifers of Texas USA. In TR Weaver and CR Lawewnce (eds.). Groundwater: Sustainable Solutions, Proc of the IAH Congress, Melborne, Australia p 1–12.

Van Tonder GJ (2001) Geohydrology Course notes, University of the Free State, PO Box 339, Bloemfontein, 9300.

CHAPTER 6

Causes and implications of the drying of Red Rock crater lakes, Australia

R. Adler and C.R. Lawrence
School of Earth Sciences, University of Melbourne, Vic., Australia

ABSTRACT: The recent drying of a group of four crater lakes in the Upper Cainozoic volcanic terrain of Western Victoria, Australia, has prompted investigation of groundwater/surface water interaction of these and neighbouring lakes. Sketchy information on the past hydrology of the crater lakes has forced several approaches to reconstruct their baseline position. Evidence supports high extraction of groundwater for irrigation from the unconfined to semi-confined aquifer of scoriacous basalt and tuff as the principal cause of groundwater depletion; whilst climate change by a reduction in the precipitation/evaporation ratio has reduced direct accession to the lakes and local recharge. Further, pre-development and current hydrologic and salt budgets of the study area indicate that there is now reversal of groundwater flow to the saline L Corangamite. Debate within the community revolves about what degree of groundwater mining that is acceptable and the merit of preserving the lakes for their environmental and tourist value.

Keywords: groundwater management, groundwater-surface water interaction, environmental value, Corangamit, Australia.

1 INTRODUCTION

The recent drying of a cluster of brackish to saline lakes of L. Werowarp, L. Gnalinegurk, L. Purdiguluc and L. Coragulac, within the Red Rock volcanic eruption complex (figure 1), prompted a hydrogeological investigation of the causes. Between 1998–2000, the lakes dried out and have not since recovered. Numerous irrigation bores tap the upper unit of the Newer Volcanics unconfined aquifer, in which the lakes lie. The salinity of the water in the lakes, 1970–1991, showed a much higher concentration (10,000–35,000 mgL^{-1} TDS) than the surrounding groundwater indicating that the lakes were at least partially of the discharge type. The water table at October 2003 was about 2.5 m below the bed of the lakes, representing a net fall in water level of approximately 5 m over an apparently short time period of about 5 years.

These crater lakes were considered to have high tourist and ecologic value, with submergent and emergent aquatic plants and habitat available for reptiles and birds. Indeed the adjacent and large L. Corangamite is recognized as a wetland of international importance under the Ramsar Convention.

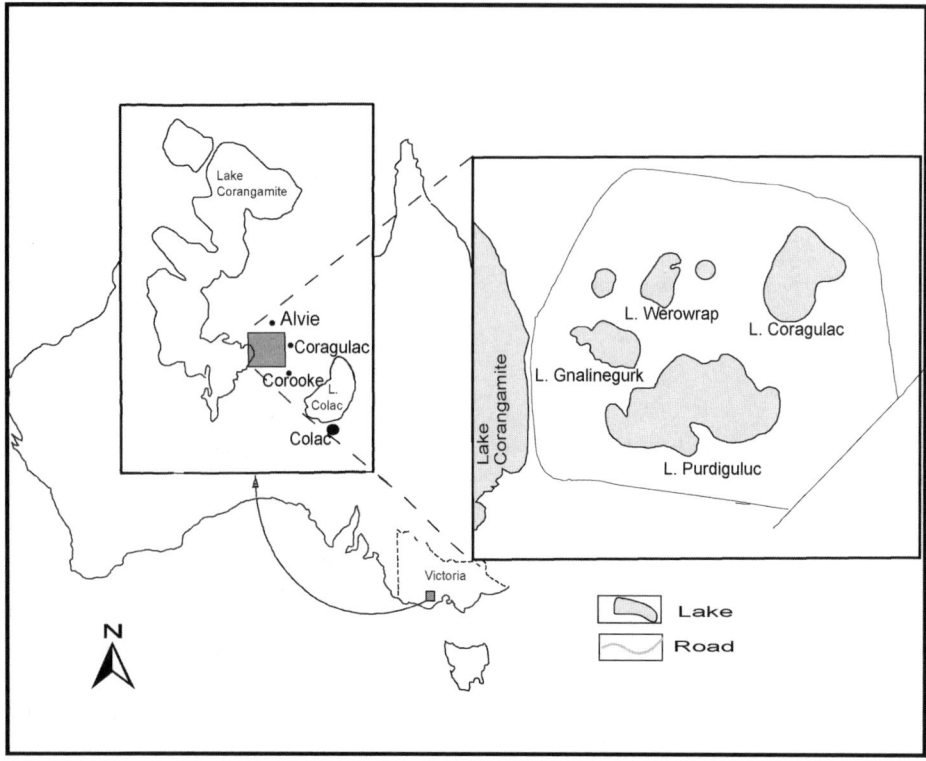

Figure 1. Locality map of the Red Rock Complex in Western Victoria.

The Newer Volcanics surrounding the Red Rocks lake complex generally have a total thickness of 30 m and an average saturated thickness of 20 m. The aquifer is heterogeneous and composed of pyroclastics and at least 2 basaltic flows with cooling joints and vesicular zones (Tickell et al., 1991). Transmissivity ranges from 10 to 700 m^2/d and the horizontal hydraulic conductivity ranges from 0.3 to 25 m/d (Gill, 1989).

The pyroclastic deposits cover the immediate area surrounding the Red Rock lakes, are highly permeable, and have developed a rich fertile soil. They contain depressions which favour recharge and associated higher quality groundwater with an average salinity of about 1,000 mgL^{-1}, TDS, which is suitable for both irrigation and stock. The bore yields from the Newer Volcanics aquifer are locally variable, with the highly vesicular zones and pyroclastic deposits having potential yields of up to 60 L/s.

Climatically the area is characterized by winter-dominated rainfall of approximately 600 mm/a and potential evaporation of approximately 1,300 mm/a. Except for the most recent lava flows or stony rises, the Newer Volcanics have been cleared of native savannah type vegetation for pasture. Despite the relatively high rainfall, because of the low relief and high hydraulic conductivity of the Newer Volcanics there is very little runoff and so have no integrated drainage system and instead there are hundreds of lakes, both perennial and ephemeral. These lakes occupy junctions between flows and collapse features in the lava. Most of the lakes are shallow (<5 m) flat bottomed with a few deep ones at eruption centres (Dahlhaus, 2004).

The investigation involved drilling and monitoring at the Red Rocks complex, assessment of the groundwater extraction, understanding the regional hydrogeology and hydrology of nearby lakes.

The water levels in lakes correspond to the groundwater table and with few exceptions there is no obvious stream inflow and no surface outlet. Investigations by previous authors (Segovia, 2001 Coram et al., 1998; Jones et al., 2001) indicate major processes contributing to lake behaviour are precipitation and evaporation. The salinity of the lakes is highly variable from $<1,000\,\mathrm{mgL}^{-1}$ TDS to $>90,000\,\mathrm{mgL}^{-1}$ TDS, controlled by the salinity of the groundwater input, evaporation and by the degree to which the lake water returns to the groundwater system.

There is debate on whether the drying of Red Rocks lakes is due to climate change or groundwater extraction. As part of this appraisal the Red Rock lakes are compared with those other nearby lakes in the volcanic plains where there is a better historical record. There is evidence that for some lakes the water level has been declining and others the level has been rising.

2 METHODOLOGY

In the absence of a reliable historical record of water levels and chemical composition of the Red Rock lakes, the limited salinity data has been used to give insight into past hydrology. Comparisons have also been made with the other lakes in the volcanic plain where there was a more complete record to help determine if the drying of the Red Rock lakes was a regional or local phenomenon.

The study focused on a prescribed study area (figure 1). A drilling program was initiated providing data on lithology, water levels and groundwater chemical composition from which current water and salt budgets were calculated. Water samples were taken from these and existing observation bores by bailing with measurement in the field of alkalinity, pH, Eh and EC, with analysis of common anions and cations in a commercial laboratory. Both the climate record and the groundwater development record were taken into account.

3 COMPARISON WITH THE HYDROLOGIC AND SALINITY RECORD OF OTHER LAKES IN THE VOLCANIC PLAINS

The hydrology and or hydrochemistry of other lakes have been studied. They include L. Murdeduke to the east, L. Bookar, L. Congulac, L. Gnarpurt to the west and northwest (Coram et al., 1998; Segovia, 2001); L. Corangamite to the immediate west (NRAEC, 1984; Williams, 1993), lakes Keilambete, Gnotuk and Bullenmerri 30 km to the west (Bowler, 1981; De Deckker, 1982; Jones et al., 2001) and Blue Lake (SA) (Radke et al., 2002). Although all lie in the Upper Cainozoic volcanic aquifer, there are some significant differences in their hydrologic behaviour.

Several lakes have been directly influenced by man's activities. In the case of L. Corangamite because it was prone to flooding affecting surrounding farmland, in 1959 a water diversion channel was constructed to redirect surface water from the Woady Yaloack river to the Barwon river and an artificial flood poundage known as Cundare Pool (NRAEC, 1984). As a result, in the period 1959–1990 the lake's water level dropped approximately 2 m and the salinity increased from 35,000 to 60,000 mgL^{-1}, TDS. For Blue Lake, Mt Gambier (SA) it has a declining water level due to extraction for the town supply (Lamontague, 2002).

For a suite of three deep volcanic crater lakes of L. Keilambete, L. Gnotuk and L. Bullenmerri, there is a persistent fall in water level since the first written records, dating from 1841. All three lakes are sub circular, saline lakes, nested in maar craters 2–4 km in diameter. Historical water levels at lakes Keilambete and Gnotuk have fallen by >15 m and at L. Bullenmerri by >20 m. The water budgets of these lakes are dominated by rainfall and evaporation and the decline is attributed to a decrease in the P/E ratio due to climate change (Jones et al., 2001), whilst the high salinity is attributed to minimal groundwater throughflow. A major difference of these lakes is that they intercept the underlying Gambier Limestone and bottom in the Gellibrand Marl aquitard.

For the widespread shallow and perennial lakes of L. Murdeduke to the east and L. Bookar, L. Colongulac L. Gnarpurt to the west and northwest (Coram et al., 1998; Segovia, 2001). L. Burrumbeet and L. Learmonth to the north. Their hydrographic record indicated some stability except for minor seasonal fluctuations and rises in wetter years such as 1983 and 1992 (Coram et al., 1998), but more recently a number of these lakes have dried out. It is thought that these lakes are the most closely allied to the historical behaviour of the lakes of the Red Rocks complex.

4 HYDROGEOLOGY OF THE RED ROCKS LAKES STUDY AREA PRIOR TO 1998

The limited historical data for the Red Rock lakes, including the 1998–2000 period when they dried up, has restricted the analysis. In assessing recorded past salinity data and lake observations an attempt has been made to understand the "natural" hydrology of these lakes prior to drying up – as a baseline position. This is complimented by an intensive field investigation, in which 10 observation bores were drilled, sampled and monitored over a 9 month period. In addition data was monitored on rainfall and groundwater extraction.

Gell (1997) recorded different salinities for the Red Rock lakes, in 1992, or in the case of L. Coragulac, 1981. In order of increasing salinity (TDS) they were: L. Coragulac (4,250 mgL^{-1}), L. Gnatinegurk (7,200 mgL^{-1}), L. Purdigulac (8,160 mgL^{-1}) and L. Werowarp (30,500 mgL^{-1}). All the lake waters were strongly alkaline, pH 9.25 to 9.7, and all characterized by high Cl, Na and Mg. According to Radke et al. (2002) the distinctive composition was strongly influenced by carbonic acid weathering of the scoria or tuff; the proportion of Na and Mg exceeds that due to cyclic sources reflecting the Na and K–rich composition of the nepheline hawaite basalt (Irving and Green, 1976).

The salinity distribution of groundwater in the study area shows an increase from east to west along the direction of regional groundwater flow from about 800 mgL^{-1} to 1,500 mgL^{-1}, TDS. The redox values indicate that all samples came from environments that have had contact with the atmosphere, as they have an Eh > 144 mV. Na$^+$ and Cl$^-$ of cyclic origin dominate the ionic composition of the groundwater of the Newer Volcanics aquifer study area, which fall into three allied composition groups: Na > Ca ≅ Mg, Cl > SO$_4$ ≅ HCO$_3$; Na > Mg > Ca, Cl > SO$_4$ > HCO$_3$; and Na > Ca > Mg Cl > HCO$_3$ > SO$_4$.

The current salinity pattern as shown in figure 2 is thought to be little changed from that prior to 1998, although a possible exception is the investigation bore alongside L. Purdiguluc in which the groundwater salinity was 7,200 µS/cm much higher than surrounding

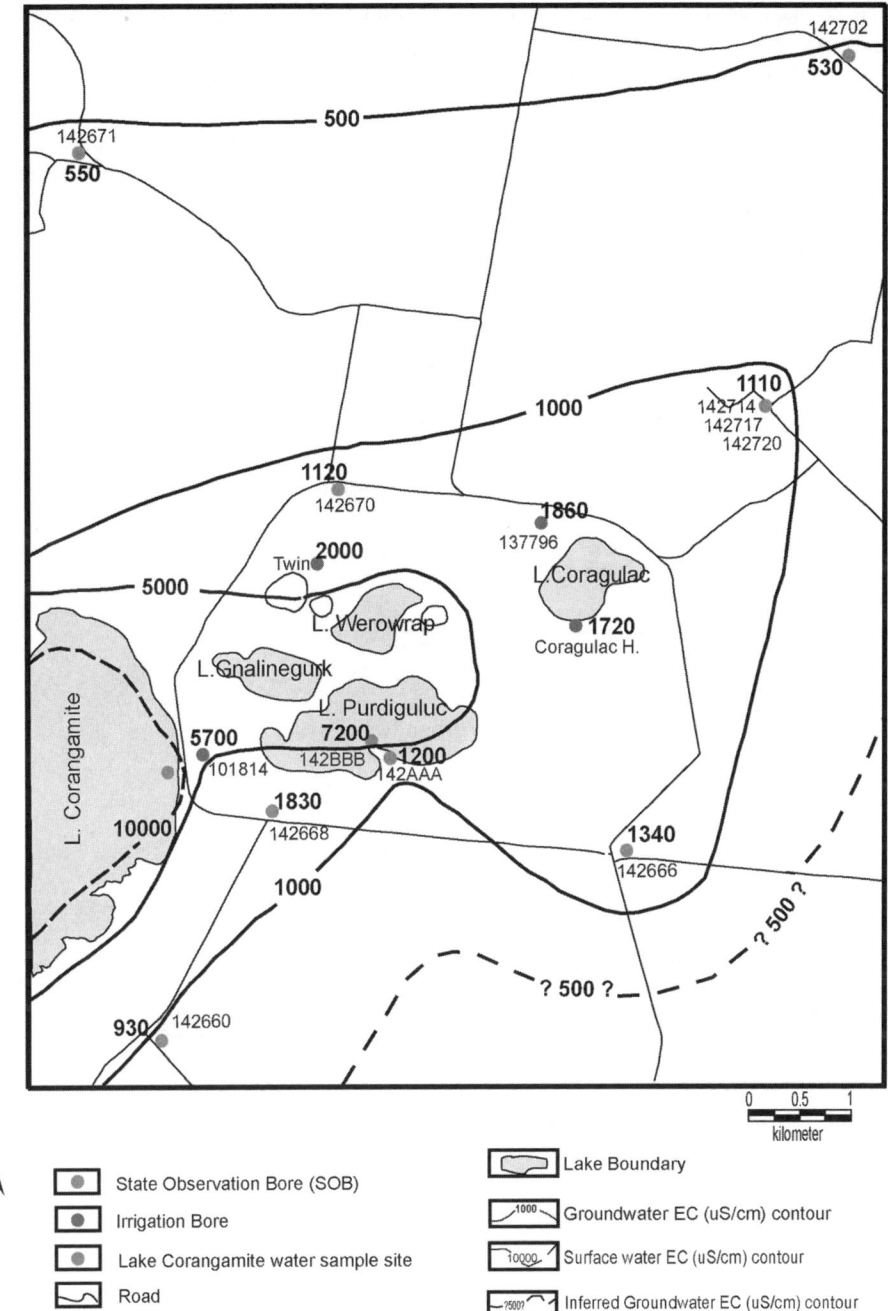

Figure 2. Salinity (µS/cm) of groundwater in the Newer Volcanics aquifer of the study area, June 2003.

groundwater. This may imply that the once saline surface water of L. Purdiguluc leaked into the groundwater system as the water table declined during the lake drying phase.

It is presumed that there was a quasi-equilibrium in terms of lake storage and salinity At nearby L. Murdeduke (Coram et al., 1998) and L. Werowarp (Walker, 1972) demonstrated that groundwater hydrology under natural conditions is largely controlled by precipitation and evaporation, whilst lake salinity is controlled by the rate of salt mass inflow from groundwater and rainwater, matched by the salt mass outflow to the groundwater system.

Superimposed on these processes there is evidence of dissolution of halite as all groundwater sampled has a Cl/Br molar ratios greater than 700, indicating a relative increase in Cl^- since recharge. This is attributed to dissolution of halite blown from the shore of L. Corangamite. There is also evidence of saline intrusion currently occurring near the eastern shore of L. Corangamite, induced by groundwater pumping, where the groundwater has similar evaporation signature to the lake water in terms of stable isotopes.

The high relief, hydraulic conductivity and evaporation rates of the study area suggest that the main influences on groundwater composition are likely to be direct accession of precipitation, rock weathering and halite dissolution, evaporation and the degree of return of lake water to the groundwater system (Coram et al., 1998).

In terms of mass balance components for the natural steady state situation these can be represented as follows: As there is no surface inflow or outflow to a typical lake in the Newer Volcanics and the salinity is assumed to be stable it follows over any given time period:

$$PC_p + G_iC_i = EC_e + G_oC_o$$

Where,
P = volume of precipitation over lakes
G_i = volume of groundwater input
E = volume of evaporation from the lake
G_o = volume of lake water rejoining the groundwater system
C_p = salinity of rainfall
C_i = salinity of groundwater input to the lake
C_e = salinity of vapour evaporating from the lake
C_o = salinity of lake water returning to the groundwater system

Then $PC_p - EC_e = G_oC_o - G_iC_i$ \hfill (1)

And if we assume that the water level under natural conditions is stable, then

$$P + G_i = E + G_o$$

$$\therefore G_i = E + G_o - P \hfill (2)$$

Substituting Eqn (2) into Eqn (1)

$$G_o = \frac{P(C_p - C_i) + E(C_i - C_e)}{(C_o - C_i)}$$

$$G_i = \frac{E(C_o - C_e) + P(C_p - C_o)}{(C_o - C_i)}$$

Thus to determine the throughflow ratio, or proportion of the groundwater inflow that becomes throughflow, to return to the groundwater flow system.

$$\frac{Go}{Gi} = \frac{P(Cp - Ci) + E(Ci - Ce)}{E(Co - Ce) + P(Cp - Co)} \qquad (3)$$

Applying Eqn (3) and using L Purdiguluc as an example:

lake salinity, $Co = 7{,}200\,\text{mgL}^{-1}$ TDS
salinity of groundwater inflow, $Ci = 1{,}000\,\text{mgL}^{-1}$ TDS
rainwater salinity, $Cp = 15\,\text{mgL}^{-1}$ TDS
salinity of vapour from lake, Ce, assumed to be 0.0005 lake water salinity = $4\,\text{mgL}^{-1}$ TDS
potential evaporation, $E = 600$ mm/a (uncorrected for salinity effect on evaporation)
rainfall, $P = 1{,}300$ mm/a
$Go/Gi = 0.14$

More saline lakes would have a lower throughflow ratio and conversely lower salinity lakes would have a higher throughflow ratio. Thus salinity information can help provide insight into the interconnection of lake hydrology with the groundwater system.

5 HYDROLOGY OF THE STUDY AREA AFTER 2000

Apart from the rainfall records and groundwater allocations for irrigation there are no monitored records on the groundwater levels and the lake levels immediately before and during the process of lake drying.

The potentiometric map (figure 3) and groundwater salinity map (figure 2) indicate that groundwater flow is generally westward and directed to the dry lakes, beneath which the water table is more than 2 m deep.

The hydrographic record for the period November 2000 to September 2003 (figure 4) indicates minor seasonal changes with the highest levels in spring and the lowest levels in autumn coinciding with the end of the pumping period. Although there is evidence of stabilization of water levels as a result of the embargo introduced in 2000 on the issue of further groundwater irrigation licences, a risk of longer term decline in water levels threatens if existing irrigation bores use their full allocation.

The relative influences of reduction in recharge because of lower rainfall and increase in groundwater extraction as a result of lower rainfall are difficult to sort out.

6 DISCUSSION ON IMPACT OF CHANGES IN RAINFALL PATTERN AND GROUNDWATER PUMPING

Possible causes of the decline in the water level of the Red Rock lakes are:

(a) Sympathetic decline in the water level due to controlled reduction of surface water inflow to the neighbouring L. Corangamite faced with an expanding L. Corangamite controls were introduced in 1959 but were not accompanied then by any obvious decline in the water levels of Red Rocks lakes.

158 *Groundwater flow understanding from local to regional scale*

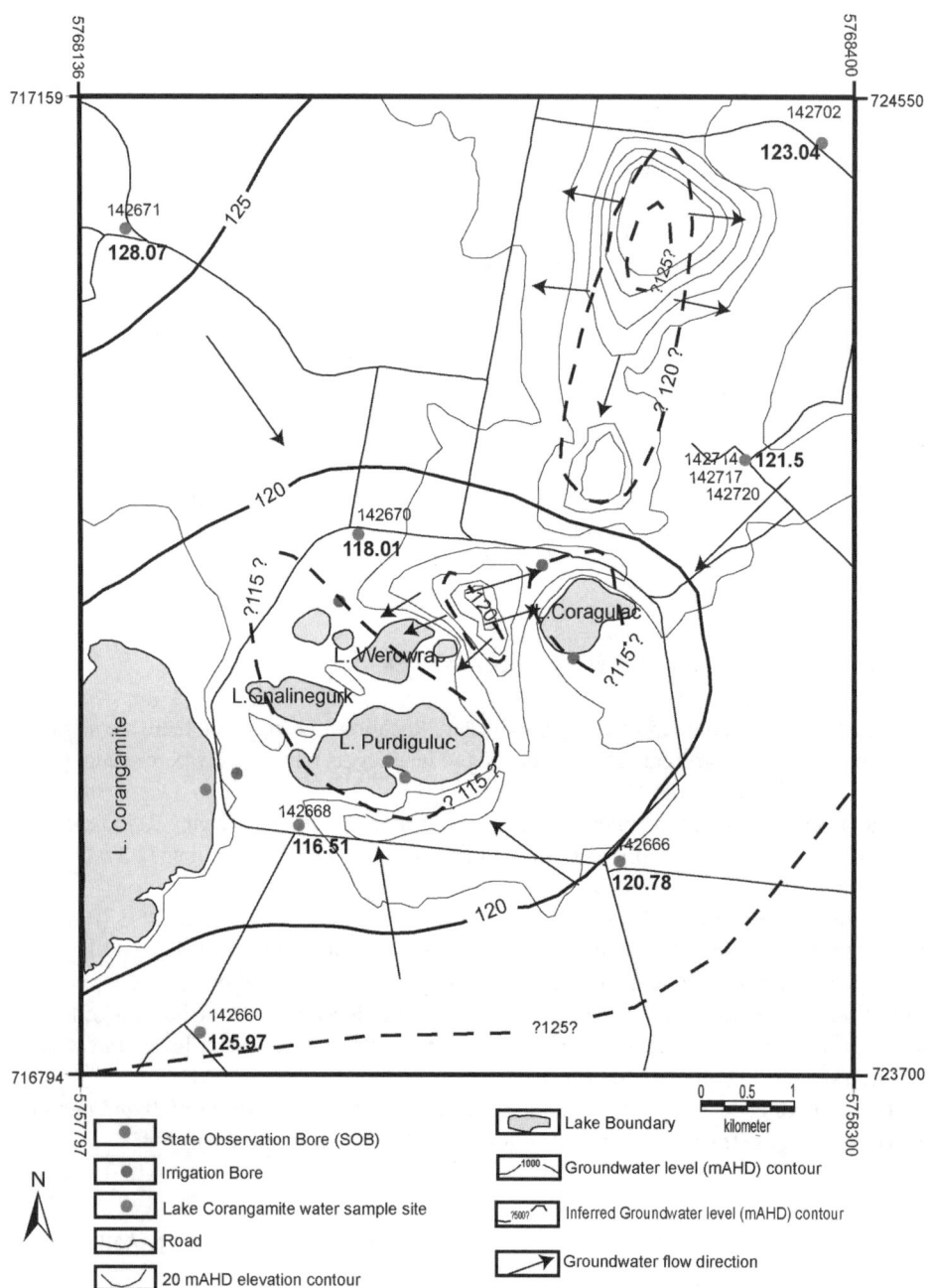

Figure 3. Potentiometric map of the Newer Volcanics aquifer in the study area.

(b) Decline in the rate of groundwater recharge and direct rainwater intake to the lakes.
Analyses for the rainfall data for the south western region of Victoria (Whetton *et al.*, 2002) indicates that over the 1990's there has been an incrase in the frequency of serious rainfall deficiency, which has continued into the 2000's. Further there has been

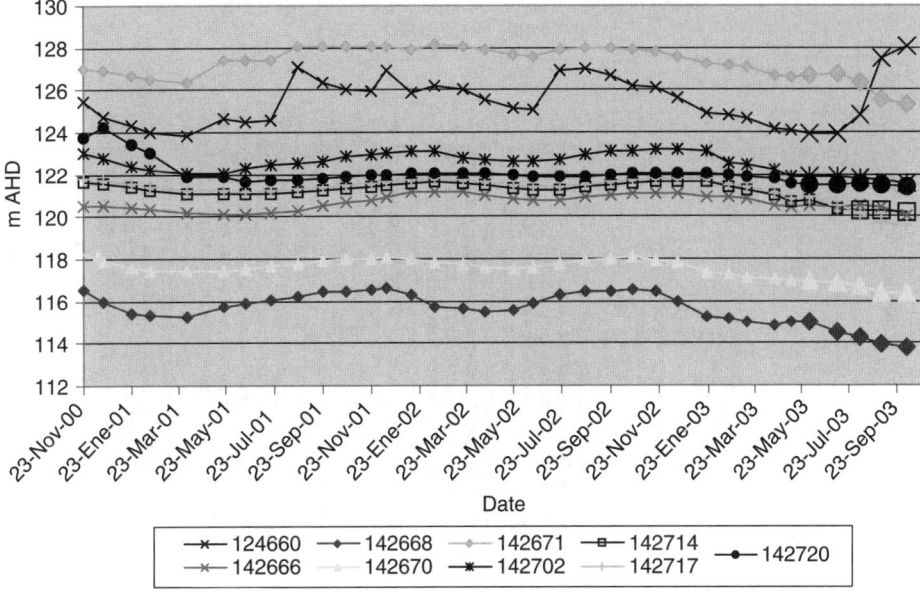

Figure 4. Observation bore hydrographs, 2000–2003.

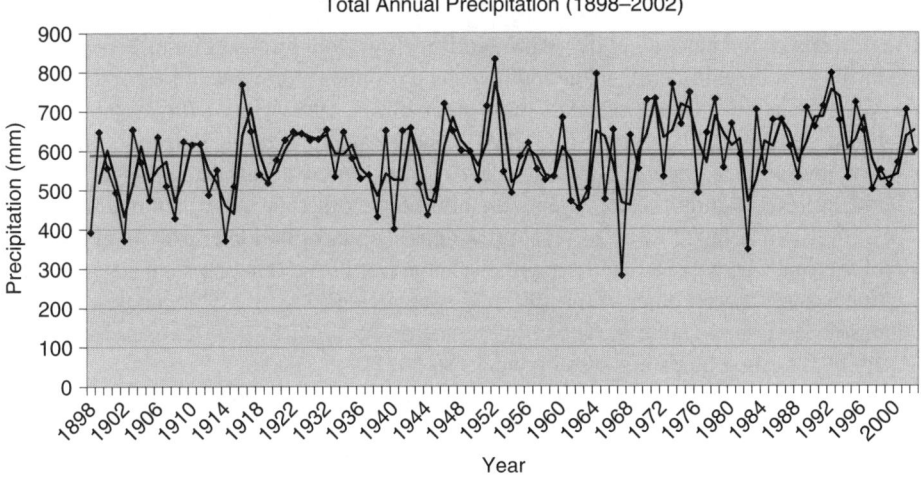

Figure 5. Monthly and annual precipitation at Warrion Hill, 1898–2002.

some decline in groundwater levels throughut the volcanic plains over the past decade and drying of shallow lakes throughout this province, as for example, Lake Learmonth in 2001 and Lake Burrumbeet in 2004.

The inference is that a major cause of drying of lakes across the volcanic plain has been a decline in recharge.

(c) Significant increase in the groundwater extracted.

Groundwater in the study area is primarily used for irrigation; the demand has greatly increased during the last 30 years, favoured by the combination of rich friable

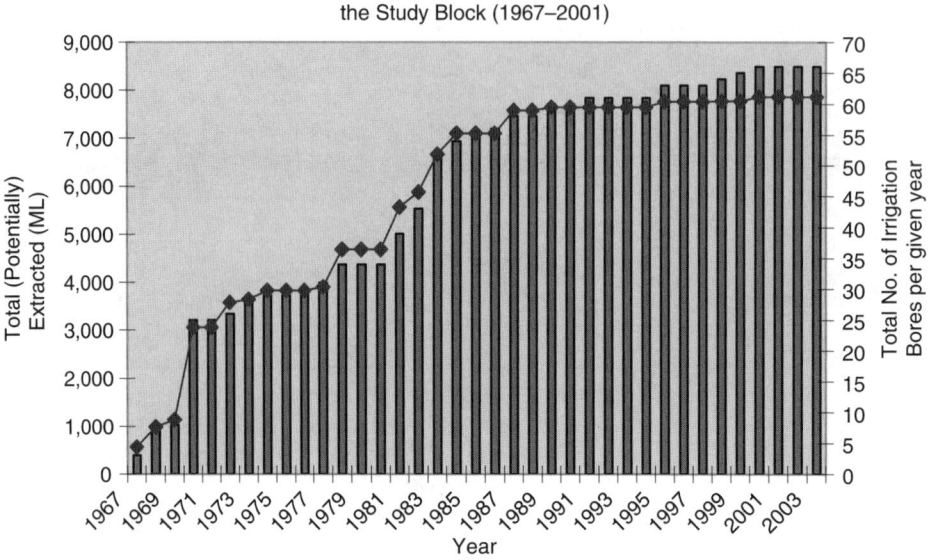

Figure 6. Number of irrigation bores and allocated extraction (ML/a or 1,000m³/a) study area 1967–2003.

soils and a conductive aquifer with good quality groundwater. Figure 6 depicts the increase in the number of irrigation bores and groundwater extracted from the Newer Volcanics aquifer for the selected study block of $8 \times 10^7 m^2$, since 1967. This graph is based on allocation values (SRW, 2003). The greatest period of irrigation bore construction occurred between 1970 and 1987, with 33 new bores. According to government records within these 17 years the allocated extraction more than doubled from $3.1 \times 10^6 m^3/a$ to $7.7 \times 10^6 m^3/a$. However based on information elsewhere in the State of Victoria where there is metering of discharge from irrigation bores the actual extraction is a lesser percentage of the allocated aggregate rates. Since 2000, no new licenses have been issued due to sustainability concerns (WGSAPCC, 2002). Currently there are 66 irrigation bores in the study block (figure 1).

The natural steady-state system had a horizontal regional groundwater flow direction from the northeast toward L. Corangamite as suggested by figure 3. The water table intersected the depressions in the Red Rock Complex and a freshwater/saline water interface was located at the eastern shore of L. Corangamite as suggested by figure 2.

7 INDICATIVE HYDROLOGIC BUDGETS FOR THE STUDY AREA – PRE AND POST DEVELOPMENT

Based on a conceptual understanding of the groundwater flow system and its interaction with the lakes, and response to development it has been possible through several analytical techniques to make estimations of the various components. Thus indicative hydrologic budgets for the study block pre and post development are given in figure 7.

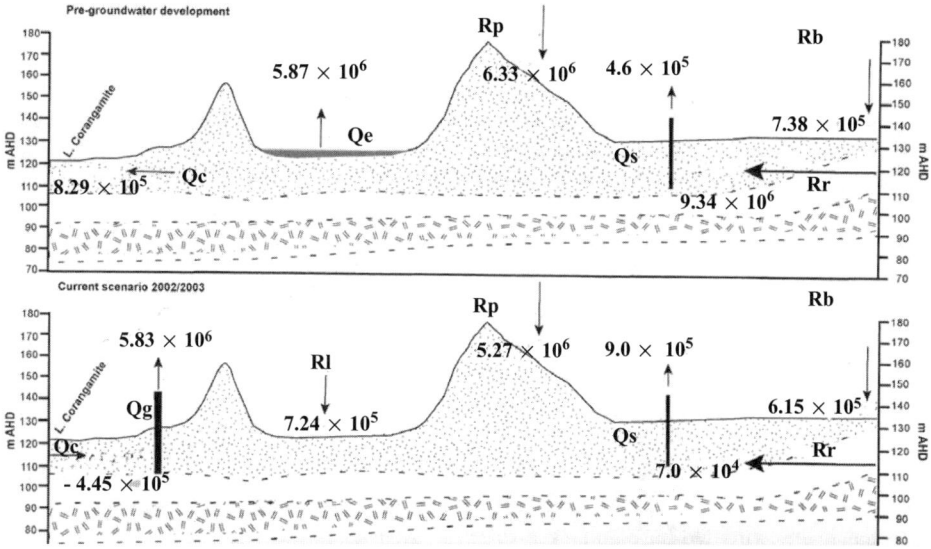

Figure 7. Preliminary annual hydrologic budgets for pre development and with development (2002/3) (units m³/a).

Principal observation is that with the lowering of the water table the Red Rock lakes have changed from primarily throughflow/discharge lakes to areas of recharge accompanied by reversal of groundwater flow to L. Corangamite.

Pre-development of groundwater for irrigation (steady state)

$$R_b + R_p + R_r = Q_e + Q_c + Q_s +/- \Delta S \qquad (4)$$

Where;
R_r = annual regional throughflow for the Newer Volcanic aquifer
R_p = annual local recharge for the Newer Volcanics (scoria)
R_b = annual local recharge for the Newer Volcanics (basalt)
Q_e = annual net evaporation from Red Rock lakes
Q_s = annual groundwater extraction from stock and domestic bores
Q_c = annual groundwater discharge to Lake Corangamite
ΔS = annual change of groundwater storage, assumed to be zero

Current scenario with development embargo

$$R_l + R_b + R_p = Q_s + Q_g + Q_c +/- \Delta S \qquad (5)$$

Where
R_l = annual recharge through lake floor
Q_g = annual groundwater extraction for irrigation
ΔS = annual change of groundwater storage, assumed to be zero

For this accounting exercise:

R_r = has been estimated from Darcy's law and assuming that the average hydraulic conductivity of the Newer Volcanics aquifer is 1 m/day.

R_p = has been estimated in part from sensitivity analysis and is assumed to be 20% of average approximate precipitation of 600 mm/a for the pre-development period and 500 mm/a post 1997 (and assuming that pyroclastic unit of the Newer Volcanics represents 75% of the study area excluding the lake area). No attempt has been made to estimate an instantaneous recharge.

R_b = has been estimated in part from sensitivity analysis and is assumed to be 3% of the average approximate precipitation of 600 mm/a for the pre development period and 500 mm/a for post 1997 (and assuming that the basalt unit of the Newer Volcanics represents 25% of the study area excluding the lake area).

Q_e = calculated from the average pan evaporation measurements at Cressy of 1,340 mm/a, corrected with a pan constant of 0.8, although no correction has been made for the salinity of the lake water and the annual average precipitation of 600 mm/a.

Q_S = estimated from the number of stock and domestic bores.

Q_c = estimated as the remaining term for the budget equations, noting that for the post-development state there has been reversal of the hydraulic gradient to L. Corangamite; the calculated value of Q_c for the post development state is therefore negative.

R_l = has been estimated based on a recharge rate of 15% of rainfall and a total lake floor area of 96,59,840 m².

Q_g = annual groundwater extraction for irrigation and is based on an assumed figure of 75% of the allocated rate, to be refined by discharge meters currently being installed.

ΔS = annual change of groundwater storage, assumed to be zero for the pre-development state, it has also been assumed for the post-development state, based on the stabilized groundwater levels over the past 2 years, to be zero.

8 MANAGEMENT IMPLICATIONS – A CONFLICT OF COMMERCIAL VALUES VERSUS ENVIRONMENTAL VALUES

Apart from advancing the conceptual understanding of the groundwater system in response to stress and the associated hydrologic accounting it is necessary in the context of sustainability to resolve the target groundwater level and extraction rates for this area. There are differing views between the stakeholders, who want a high groundwater extraction for their enterprises, and those who value the natural environment highly, wishing the lakes to be restored to a lake-full state supporting a rich ecology of submergent and emergent aquatic plants, water birds and reptiles (LCC, 1976; Robertson, 1998) and provide a spectacular tourists attraction.

Given the sustainability and environmental precepts of the Water Act 1985, new monitored data on metered volumes from irrigation bores, water levels of observation bores and community input it is possible that some reduction in extraction rates will be required. This would need to be staged, with the center pivot/bore systems nearest the lakes of higher priority.

9 CONCLUSIONS

- There is strong evidence that increasing groundwater extraction for irrigation over the past 30 years has depleted the groundwater resource and dried the lakes. Salty water that occupied the lakes has now entered the groundwater system and because of this depletion and from the large L Corangamite is continuing to do so. Also there is evidence that

drier climate conditions in south-western Victoria over at least the past decade have led to drying of other lakes across the volcanic plains, even where there is not intensive groundwater extraction nearby. It is concluded that drying of Red Rock lakes reflects both the localized impact of high groundwater extraction and the regional impact of prolonged drought conditions.
- There are signs that the embargo on further allocation of groundwater extraction licences has helped arrest the decline in the water table. However the lakes still remain dry as of August 2005 and a reduction in the actual groundwater extraction rate may be required if the lakes are to be restored.
- There is concern that the Red Rocks case is a signal of climate change with implications for management of watr resources in south–eastern Australial.

ACKNOWLEDGEMENTS

The authors wish to thank the Department of Sustainability and Environment who provided the grant for this study and Gordon Walker (DSE) and Ass Prof B Joyce of the School of Earth Sciences, University of Melbourne for comment.

REFERENCES

Bowler, J.M (1981) Australian salt lakes: a paleohydrologic approach. Hydrobiologia. 82: pp. 431–444.
Coram, J.E., Weaver, T.R. and Lawrence C.R (1998) Groundwater- surface water interactions around shallow lakes of the Western District Plains, Victoria, In Proc. IAH Groundwater Conference, Feb 1998: 119–124.
Dahlhaus, P (2004) Characterising groundwater flow systems for salinity management in the Corangamite Region, Australia. Proc IAH ALSUD Congress, Zacateccas, Mexico 2004.
De Deckker, P (1982) Holocene ostracods, other invertebrates and fish remains from cores of four maar lakes in southeastern Australia. Proc. Roy. Soc. Vict. 94, pp. 183–219.
Gell, P.A (1997) The development of a diatom distribution for inferring lake salinity, Western Victoria, Australia: towards a constructive approach of reconstructing past climates. *Aust J Bot* 45: 389–423.
Gill, B (1989) Hydrogeological Review of Salinity Problems in the Barwon/Corangamite Region. Investigations Branch No. 1988/16. Rural Water Commission of Victoria.
Irving, A.J. and Green, D.H (1976) Geochemistry and petrogenesis of the Newer Volcanics. *J. Geol Soc Aust.* 23: 45–66.
Jones, R.N., McMahon, T.A. and Bowler, J.M (2001) Modelling historical lake levels and recent climate change at three closed lakes, Western Victoria, Australia (c.1840–1990) >Jour of Hydrology 246:159–180.
Joyce, E.B (1988) Cainozoic volcanism in Victoria, Chapter 8, in Clarke, I. and Cook, B. (eds) In Victorian Geology Excursion Guide, Australian Academy of Science, Canberra, pp. 71–80.
Land Conservation Council (LCC) 1976. Report on the Corangamite Study Area. Melbourne.
Robertson, P (1998) Corangamite Water Skink: National Recovery Plan 1998–2003, Environ Australia Biodiversity Group.
Lamontagne, S (2002) Groundwater delivery rate of nitrate and predicted change in nitrate concentration in Blue lake, South Australia. Marine and Freshwater Research. 53(7): 1129–1142.
Natural Resources and Environment Committee (NRAEC) 1984. Stage One Augmentation of Geelong's Water Supply to the year (1995). Background Information Paper. 90p.
Radke, L.C., Howard, K.W.F. and Gell, P.A (2002) Chemical diversity in south-eastern Australian saline lakes 1: geochemical causes. Marine and Freshwater Research 53: 941–959.

Segovia, I (2001) A study of nutrients in Lake Murdeduke in Western Victoria and implications for environmental MSc thesis. University of Melbourne.

Southern Rural Water (SRW) 2003. Groundwater Management Plan. Report for the year ended 2003.

Tickell, S.J., Cummings S., Leonard J. G. and Withers J. A (1991) Colac Geological Report, 1: 50,000 map. Geol Surv Report No. 89.

Walker, K.F (1972) Studies of a saline lake ecosystem. *Marine and Freshwater Res 24: 21–71*.

Williams, W.D (1993) Australian inland waters: A limited resource. Australian Biologist 6: 2–10.

Warrion Groundwater Supply Protection Area Consultative Committee (WGSPACC) 2002. Warrion Groundwater Supply Protection Area: Explanatory Notes to the Warrion Groundwater Management Plan.

Water Act (1989) Victorian Government Printer, Camberra, Australia.

Whetton, P.H., Suppiah, R., McInnes, K.L., Hennessy, K.J., and R.N. Jones (2002) Climate change in Victoria. High resolution regional assessment of climate change impacts. Department of Natural Resources and Environment Victoria 48p.

CHAPTER 7

The development of a methodology for groundwater management in dolomitic terrains of South Africa

S. Veltman[1] and B.H. Usher[2]
[1] The Department of Water Affairs and Forestry, Bloemfontein, South Africa
[2] The Institute of Groundwater Studies, University of the Free State, Bloemfontein, South Africa

ABSTRACT: In this paper the authors make use of the *PCME* (prior conceptual model explanation) approach, to develop a technical groundwater management methodology and a first order technical groundwater management tool for dolomitic terrains in South Africa. This will enable water managers to manage groundwater hydrogeologically in dolomitic compartments, with the focus on volumes available in the aquifer for future allocations. The principles of Integrated Catchment Management (ICM) and Integrated Water Resource Management (IWRM) are an integral part of the methodology. The aim is a practical methodology, which can be altered as new data and information becomes available, rather than an exhaustive methodology. The methodology was developed partially from information attained in a case study of the Schoonspruit dolomitic compartment, during which the general aquifer characteristics were explored, and the groundwater flow system regime defined.

Keywords: arid regions, groundwater budget, groundwater management, karst, South Africa.

1 INTRODUCTION

> "All capable hydrogeologists use past experiences, principles, generalisations and qualitative linguistic modelling, applying this to systematic hydrogeological reasoning. This is referred to as prior conceptual model explanation (PCME) and represents an initial high grade, synergistic analyses of hydrogeological foreknowledge, derived largely from existing information. The objectives are: (1) to acquire optimal value from existing information, (2) to reach a high level of knowledge as a basis for further study, and (3) to provide an early perspective to be explained to involved stakeholders".
>
> (LeGrand and Rosen, 2000)

The Department of Water Affairs and Forestry (DWAF) is responsible for managing the quantity and quality of water resources. Changes to the executive framework of water resource management have given regional offices of the DWAF responsibility of managing local water resources. This responsibility renders a new commitment with regard to water resource management. Regional offices have to make decisions, based on sound scientific principles, as to allocable water resources.

For planners and managers dealing with groundwater and in particular dolomitic aquifers, to fulfil their purpose, a consistent practical methodology was deemed necessary, since no official methodology existed for groundwater management in dolomitic terrains. The challenge therefore was to develop and test an appropriate tool and use it as a practical technical groundwater management tool. The *PCME* approach was decided on and used to develop and test the methodology and develop a practical technical groundwater management tool for the Schoonspruit dolomitic compartment (dolomitic compartment is a section of original dolomite that has been compartmentalised with impermeable geological features such as dolerite dykes).

The deliverables of this project include: (1) an adaptable and workable methodology, and (2) a tool to use for allocation of groundwater extraction volumes in the dolomitic compartments.

2 THE SCHOONSPRUIT DOLOMITIC COMPARTMENT

Information collated for the *Hydrogeological Evaluation* of the Schoonspruit dolomitic compartment included the topography, vegetation, meteorology, drainage, geological and hydrogeological features, water users and legal proclamations. Desktop information was evaluated and calibrated against previous work and where necessary, additional fieldwork was carried out. Additional fieldwork included a précised surveyed hydrocensus of groundwater features. Through the collation and evaluation of the information a better understanding of the aquifer characteristics, the aquifer domain and the flow system regime could be achieved on the local and regional scale.

This case study as an example show the information and process necessary for effective management of the dolomitic system.

2.1 Geographical features

The Schoonspruit dolomitic aquifer is situated, figure 1, to the North and Northwest of the town Ventersdorp in the Northwest Province, South Africa. The topography slopes downward from the Northeast to the Southwest with elevation changes of more about 100 m over a 40 km distance and circular depressions can be found in the area that shows elements of karstic evolution. The area has summer rainfall with most of the precipitation occurring from October to April and is drained by the Schoonspruit River.

2.2 Geological features

The dolomitic compartment is formed by rocks of the Transvaal Sequence, figure 2, Kotze (1994) and includes:

- The Black Reef Formation to the south of the compartment, mainly quartzite that is less than 1 m thick.
- The Chuniespoort Group dolomites, which was part of a chemical sedimentation phase with the Malmani Subgroup representing the main dolomitic stage in the chemical sedimentation phase. The Malmani Subgroup can be up to 1,550 m thick.
- The Pretoria Group, quartzite and shale, which comprises of the Rooihoogte Formation in the north of the compartment and is only of significance because of the boundary effects and runoff generated from this area towards the dolomitic compartment.

The development of a methodology for groundwater management in dolomitic terrains 167

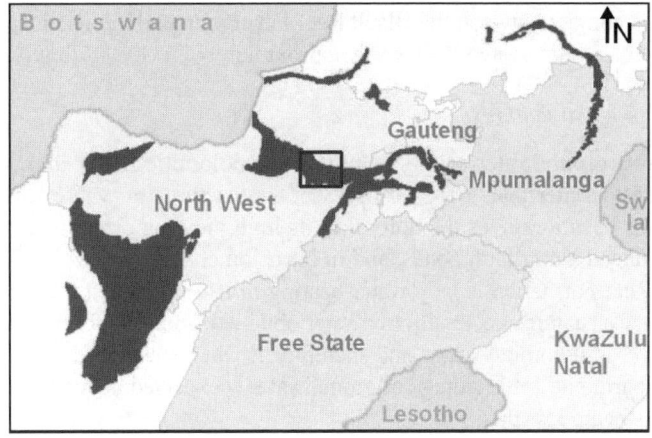

Figure 1. Location of the Schoonspruit dolomitic compartment in South Africa.

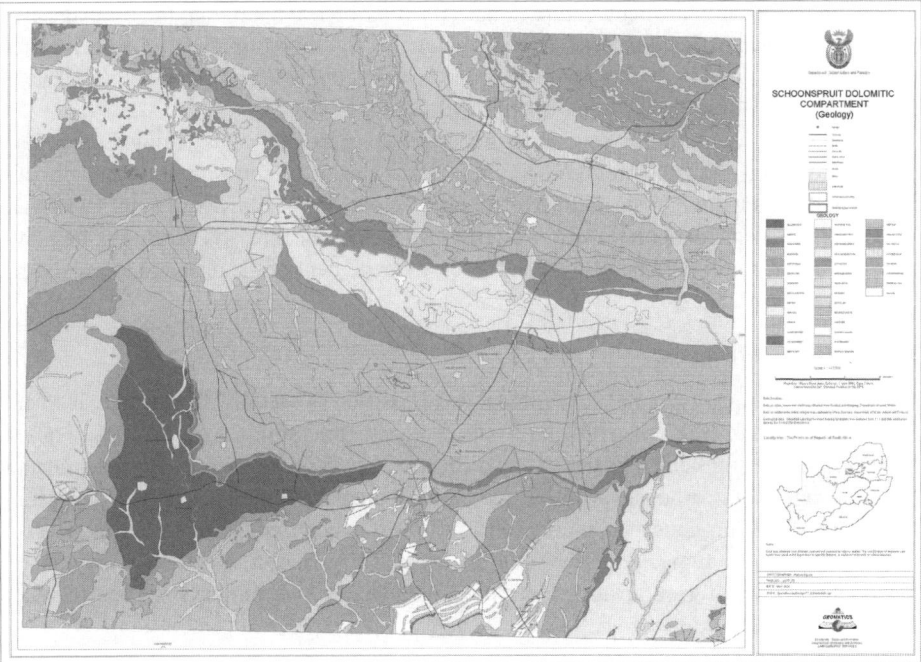

Figure 2. General geology of the Schoonspruit dolomitic compartment.

Two main trends of dykes exist, NNW-SSE and WSW-ENE. From the groundwater flow perspective the importance of dykes is twofold: (1) they form preferential pathways and (2) they can act as flow boundaries. The boundaries of the Schoonspruit compartment have formed as a result of the difference in geological formations and structural control; they may be defined as:

- North – The Pretoria Group and Blaauwbank Dyke
- East – The NNW-SSE dyke following approximately 27° longitude

- South – The contact between the Black Reef Formation and Ventersdorp Supergroup
- West – The N-S fault system following approximately 26°30′00″ longitude.

2.3 Hydrogeological features

In the study area groundwater occurs mainly in the dolomitic rocks; high yielding boreholes, producing greater than 10 l/s, are present at structures or where karstification has developed. Karstification gives the dolomites its high yielding properties and dissolution is more pronounced along fault zones, and in intrusion contact zones.

The Black Reef Formation is only water bearing in its upper-most weathered zones and where secondary structures occur. Groundwater obtained from this formation has the same chemical nature as dolomitic water and therefore it can be assumed to be linked to the dolomitic compartment. Little storage of groundwater is expected in this formation, because of the thickness being less than 1 m.

The Malmani Subgroup is the most productive and sustainable aquifer and extraction estimated at 35 Mm^3/a. Borehole yields range from 3–12 l/s, with chert-rich layers yielding 11–12 l/s.

The dolomitic compartment, figure 2, has an extent of 1,585 km^2 and the drainage area of a spring known as Schoonspruit Eye is 840 km^2. A flat topography and fast infiltration rate results in low flow contribution to surface water bodies. The Schoonspruit Eye is dependent on the dolomitic compartment for flow.

Four separate communities, several farmers (including irrigation) and mining operations are dependent on the dolomitic compartment for water, extracting groundwater directly from boreholes situated on the compartment. The Ventersdorp Municipality and two irrigation boards are dependent on flow of springs for surface water flow and executing their lawful water use. Aquifer parameters relevant to the management of the resource are described further in section 2.6.

2.4 Water quality

The groundwater quality in the compartment was classified as typically hard to very hard and moderately alkaline, with a total dissolved solid content ranging from 200 to 748 mgL^{-1} or less than 150 mS/m (milliSiemen/metre) of electrical conductivity. Groundwater is predominantly calcium-magnesium-carbonate with Mg/Ca ratio ranging between 1.2 and 1.8, suggesting recently recharged groundwater and typically infiltrated in dolomitic terrain (Polivka, 1987).

Nitrate levels are typically, in the order of 2 mgL^{-1}. Agricultural activities may pose a risk on the quality of groundwater, therefore a quality early warning system was included in the *Groundwater Management Tool*.

2.5 Water levels

The correlation between surface topography and groundwater level in the dolomitic compartment is 51.8%, figure 3, indicating that the surface topography does not govern the water level elevations in the dolomitic compartment and therefore, Bayesian interpolation could not be used for the contouring of the spatial distribution. Kriging interpolation was used to interpolate water levels, figure 4, for use in spatial groundwater level contour maps and to determine flow directions. The groundwater gradient at the Schoonspruit Eye is very flat and any extraction will influence the flow direction of groundwater. This is observed throughout the compartment, since changes in the water table has a substantial effect on the groundwater flow vectors.

The development of a methodology for groundwater management in dolomitic terrains   169

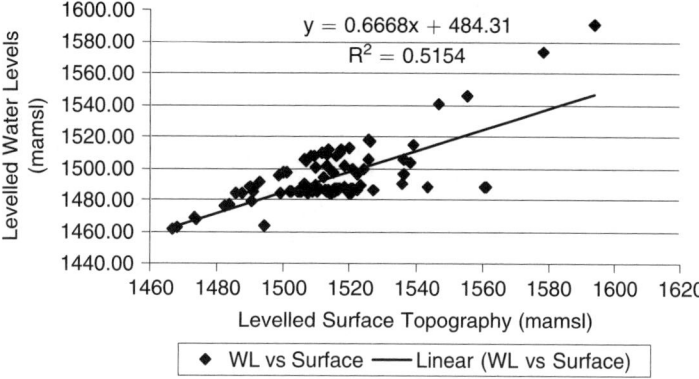

Figure 3. Correlation between water level and surface topography elevations.

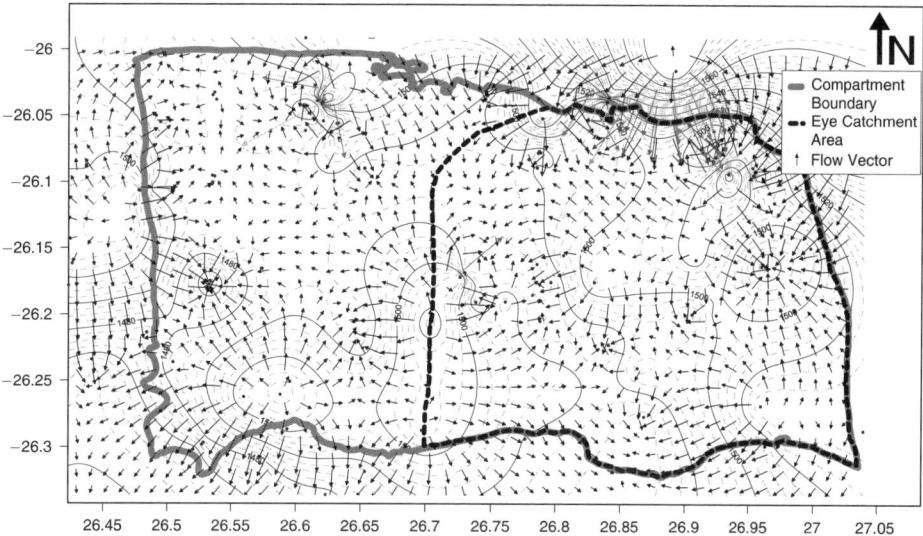

Figure 4. Water level contours and flow vectors in the Schoonspruit compartment.

It is assumed that limited inflow of groundwater occurs through the eastern and northern boundaries and outflow occurs through the southern (0.245 Mm^3/a) and south-western (0.391 Mm^3/a) boundaries. Regarding the volume of water stored in the dolomitic compartment, (Polivka, 1987), it was calculated at 1,440 Mm^3 and this value was used in water balance simulations.

The delineation of the Schoonspruit Eye catchment area was drawn as a result of the flow vectors from the Kriging interpolation and knowledge of the underlying geology, figure 4. The delineation showed two groundwater management zones in the compartment; *Zone A* – the Eastern Dolomitic Eye Catchment and *Zone B* – the Western Dolomitic Compartment. Refer to section 3.4 for the topocadastral map included in the *Groundwater Management Tool* showing the two groundwater management zones.

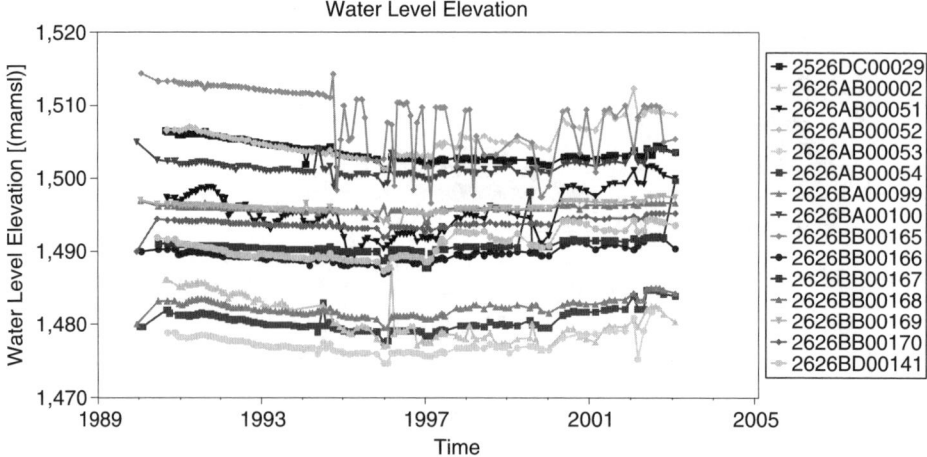

Figure 5. Water level elevation fluctuations within the Schoonspruit compartment.

2.6 Aquifer parameters

The transmissivity of dolomites is typically higher than that of the surrounding rocks and this gives rise to springs on contact zones. Transmissivity and storativity values determined from individual boreholes are of limited regional applicability, due to the karstic nature of the aquifer. Groundwater levels and rainfall data are a critical input into most simulations and evaluations, of which the Saturated Volume Fluctuation (SVF), Cumulative Rainfall Departure (CRD) and Moving Average (MA) methods give the best simulations for determining natural water levels and various aquifer parameters. Storativity values determined from the CRD and MA methods are integrated over the aquifer and therefore can be used in other simulations or regional numerical models.

Recharge in the area is high due to soils that are transmissive and areas of karstification, which allow rapid recharge, figure 5. Groundwater level time series shows that the aquifer has gained significant amounts of recharge, since water levels are at the same depth as in 1990, recovering since 1996, figure 5. Different recharge estimations were used and a level of certainty was assigned to each method, based on the certainty of the input parameters and the certainty of the applicability of the method to the aquifer, table 1.

A weighted approach was also adopted, where recharge estimates in the same ranges have a high weight and the outliers have a lower weight, although they still have an input into the recharge calculations. The hydrogen isotope displacement method yielded very little recharge for a dolomitic aquifer and the EARTH model simulation did not attain a very good fit, therefore a weight of 1 was assigned to these methods. The Qualified Guess and the Equal Volume (EV) methods was assigned a 3, because expert knowledge has been put into these methods, so it cannot be disregarded, but the evaluations on the Qualified Guesses might be outdated and the EV could only be fit for a portion of the time series data. The chloride, SVF and CRD methods have proven to be very good tools for determining recharge and a weight of 4 have been assigned to these.

Recharge in the Schoonspruit Dolomitic Compartment was estimated as 6.0% of annual rainfall, amounting to 71 Mm^3/a.

Table 1. Summary of recharge estimates from various methods.

Method	mm/a	% of rainfall	Certainty (Very High = 5; Low = 1)
Cl – Zone A	182.7	29.47	4
Cl – Zone B	58.65	9.46	4
SVF: Equal Volume	30.3	4.9	4
SVF: Fit	34.1	5.5	4
CRD	35.4	5.7	4
^{2}H displacement method	16	2.6	1
EARTH Model	37.2	6.0	1
Qualified Guess	43.4	7.0	3
Average recharge	37.2	6.0	

mm/a: millimetres per annum; Cl: Chloride; SVF: Saturated volume fluctuation; CRD: Cumulative rainfall departure; ^{2}H: ydrogen isotope

2.7 Dolomitic Springs

According to Polivka (1987) the compartment recharges six springs in the area of which the Schoonspruit Eye is the most prominent and surfaces in an approximate area of 5 km². The flow of the eye can be simulated using various methods of which the MA method has proven to be the most effective. When the moving average of rainfall is used instead of true rainfall figures one can incorporate the lag time effect of rainfall events and its integration over the aquifer, Bredenkamp et al. (1995).

The recharge relationship of the spring flow to various monthly averages of rainfall was defined and is vital for good groundwater management of the Schoonspruit Eye. Polivka (1987) calculated that the dolomitic compartment needs at least an annual rainfall above 313 mm to be recharged, and even then only 30% of the rainfall in excess of this value contributes to the annual recharge. This value was determined through calibration of earlier MA simulations and used as the first input value into the MA simulations of the Schoonspruit Eye with Equation 1.

Spring flow parameters have been incorporated into the spring flow simulation together with the different moving averages of rainfall that affects the flow. Various factors have been introduced to simulate different situations, e.g. spatial extent of the drainage to the spring, extraction from the system or different lag time effects. When doing the simulation, all known parameters are incorporated and the unknown parameters are calibrated to attain the best fit for the spring flow. The Schoonspruit Eye can be simulated (figure 6) with Equation 1. This simulates the relation between spring flow and rainfall using moving averages, with different lag time effects, to account for flow through the aquifer, taking into account current impacts on the aquifer by introducing extraction from the drainage area. Too little information currently exist to determine the inflows and outflows in the drainage area of the spring and these are assumed to be insignificant in the simulation when compared to (the high) recharge from rainfall.

Schoonspruit Flow (Mm³/m) = ReN + ReF − $ExtGW$ (1)

Where

ReN	recharge under normal rainfall events (Mm³/m)
ReF	recharge under flood rainfall events (Mm³/m)
AbsGW	groundwater extraction from the drainage area (Mm³/m)

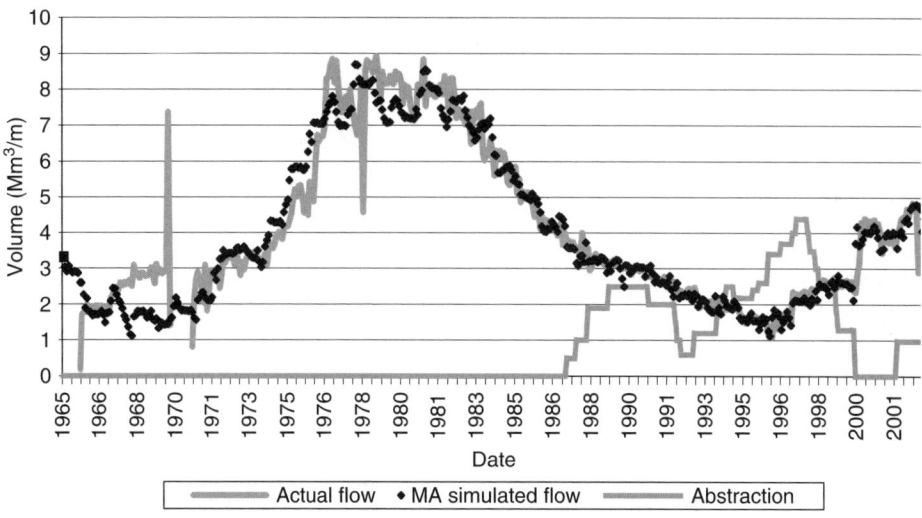

Figure 6. Schoonspruit Eye simulated flow with a 96 Month Moving Average.

And

$ReN = ReN\%/100 * Rf24MMA/Rf120MMA * (Rf96MMA - ThN) * A/1000$
$ReF = ReF\%/100 * (RfFLOOD) * A/1000$
$RfFLOOD = IF((Rf120MMA - ThF) > 0, \quad Rf120MMA - ThF)$

The calibrated parameters for the system have been determined as $ReN\%$ (7), $ReF\%$ (44), ThN (26 mm), which is the recharge threshold, and ThF (43 mm), which is the flood recharge threshold. A refers to the drainage area (842 km^2) and Rf to the month lag time included.

3 TECHNICAL GROUNDWATER MANAGEMENT METHODOLOGY

A *Technical Methodology* was developed by simplifying the hydrogeological evaluation processes and water managers and planners can simply use a flowchart as a checklist, figure 7, to identify the necessary information to successfully implement groundwater management. The various aspects of the *Technical Methodology* are discussed briefly in the sections that follow.

3.1 *Management principles*

Water resource management is based on principles of equity, optimal use, sustainable use and Integrated Water Resource Management (IWRM). This is a philosophy of co-ordinated management of an area's water, land and other resources, to maximise the resultant economic and social welfare in an equitable manner without compromising the sustainability of vital ecosystems. Therefore, it is no longer acceptable to manage groundwater in a separate manner; management of groundwater has to comply with the policy, strategy and practice of general water resource management in South Africa (Parsons et al., 2001).

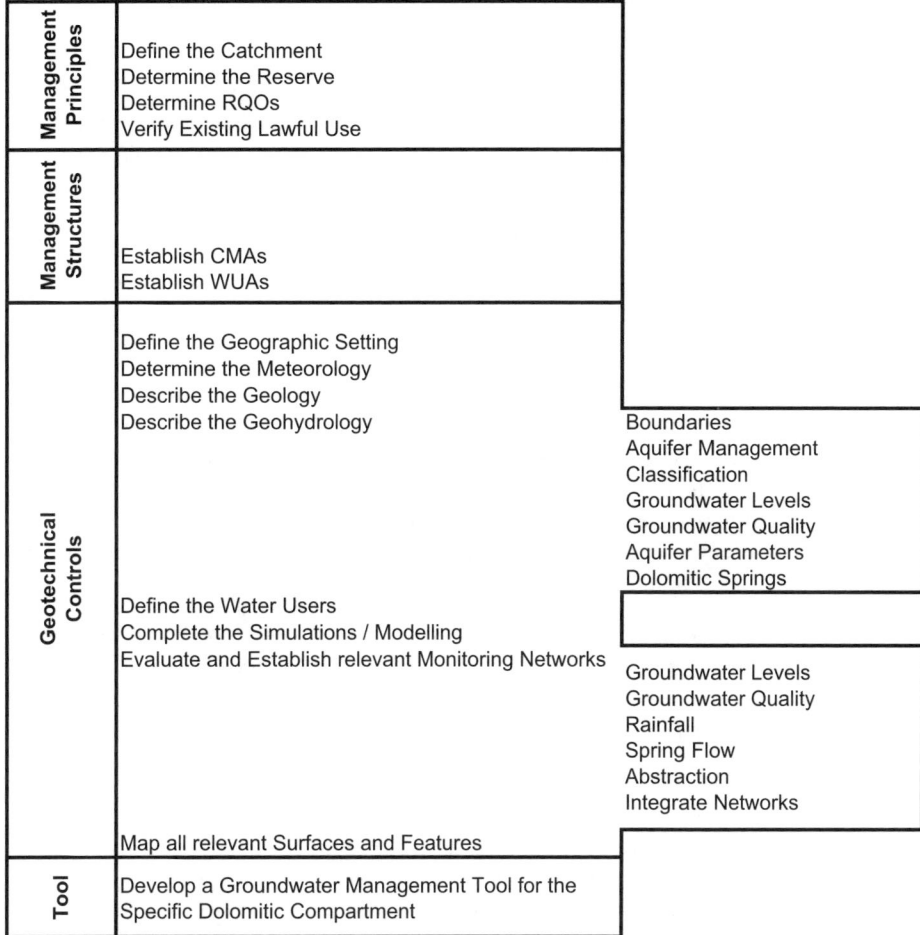

Figure 7. Steps to follow in other dolomitic areas.

Integrated Catchment Management (ICM) is a process and an implementation strategy to achieve a sustainable balance between utilisation and protection of all environmental resources in a catchment. A catchment can be defined as a physiographical drainage area determined by the runoff or recharge to the water resource, at a particular location. Naturally occurring water can only be managed effectively and efficiently within catchment or river basin boundaries, because of the need to technically account for all aspects of the hydrological cycle, as well as human interference, therefore:

- The catchment area of the dolomitic compartment needs to be defined and explained.

The National Water Act (NWA), 1998, introduces source and resource based tools to protect and conserve groundwater and they are related to both quality and quantity. Source based tools include licensing and authorisation, while resource based measures include Resource Directed Measures and classification. Resource Directed Measures are described by the NWA (1998) for protection of the water resource and consist of (Parsons et al., 2001):

- The Reserve for a water resource that needs to be determined and is defined by the NWA (1998) as two components; (1) the water necessary for basic human needs, and (2) the water necessary to sustain the ecosystem.
- Resource Quality Objectives (RQOs) that need to be set and are a numerical or descriptive statement of the conditions, which should be met in the receiving water resource, in terms of resource quality, in order to ensure that the water resource is protected. Resource quality includes four components: water quality requirements, quantity requirements, habitat integrity and objectives for biotic integrity (MacKay, 1999).

The NWA (1998) and the National Water Amendment Act (Act No. 45 of 1999), describe the legality of water uses and the types of use that need to be registered and licensed. Currently all water uses must be registered, irrespective of the legality of the water use, therefore registering a water use does not make it a lawful water use. Licensing takes place were a water user wants to use water after 1 October 1998, or where a use was registered and verified, according to Section 35 of the NWA (1998) to be unlawful.

The following check was introduced in the methodology to ensure that these management issues have been captured in an evaluation of the dolomitic aquifer.

- The legality of all water uses in the area needs to be verified and subsequent registration and licensing of the water users must be a priority and performed as soon as possible.

3.2 Management structures

A key element of water resource management is the establishment of Catchment Management Agencies (CMAs) that is responsible for managing water resources in a specific catchment. The nature of the dolomite aquifer however, requires groundwater management to remain essentially at local level and includes monitoring and control of groundwater extraction, groundwater level fluctuation, groundwater quality change, rainfall recharge and environmental impacts (Parsons et al., 2001). The NWA (1998) provides for the establishment of Water User Associations (WUAs) to enable this local level the management of dolomitic aquifers.

- CMAs must be established as soon as possible.
- WUAs must be established for each of the smallest manageable dolomitic aquifer unit.

3.3 Geotechnical controls

Various geotechnical controls exist which govern the flow regime and which can be defined or determined. The following checklist, aims to provide general instructions in a systematic way, outlining the necessary steps to follow in a hydrogeological evaluation to enable meaningful resource management.

The geographic setting of an area is easy to define and the following must be included:

- Province and nearest town. The country can be included if relevant.
- Local authority boundaries if relevant.
- Topocadastral map or other relevant maps, e.g. 1:50,000 or 1:250,000
- Geomorphological features that might be important, e.g. sinkholes.
- Type of land cover.

- Relevant catchments within which the area falls.
- Relevant surface water resources that can be used for orientation.

A good understanding of the climatic conditions of an area is essential, especially with regard to groundwater in the dolomitic areas, since rainfall is the source from which recharge originates. The following must be completed for flow simulations to be done:

- Type of rainfall season typical to the area including average rainfall and evaporation.
- Minimum and maximum temperatures.
- Mean annual precipitation (MAP), and where possible rainfall zones of similar precipitation can be compiled.
- A reliable set of monthly rainfall data, with at least 10 years of rainfall data prior to the first groundwater level or spring flow data.

The geological environment is the governing factor in how the groundwater flow system will respond. Different rocks have different water bearing capabilities and geomorphologic and structural features determine flow regimes. The following information is essential:

- Geological description according to geological maps available.
- Geological description from previous reports, including structural features and borehole logs available, focussing on possible geological boundaries.
- Geological map compiled from available detail information in reports or electronic form.
- Geological cross-section, where enough spatial information is available and where it is necessary.

The hydrogeology of the area should be described in general to attain an understanding of the groundwater system and the following checklist should be used:

- Hydrogeological description according to hydrogeological maps available.
- Hydrogeological description from previous reports including structural controls and data available.
- Hydrogeological map compiled from available detail information in reports or electronic form.
- Boundaries, geological structural features such as dykes, geological contacts and differences in geological layers in the dolomite must be described, and included in the hydrogeological map.
- The dolomitic aquifer has to be classified in terms of vulnerability (to contamination) and importance before one can decide how much effort should go into the management of the specific dolomitic area.

Groundwater level interpretations are essential to the understanding of the flow system, both on a regional and local scale, and the following actions must be performed:

- Evaluate existing time series data to determine rainfall/water level fluctuation interactions, as well as aquifer parameters.
- Plot borehole elevations and water level elevations on a 2-dimensional chart, to determine if a correlation exist between groundwater levels and the topographic surface. This is used to determine what kind of interpolation should be used for the contouring of groundwater levels and to define recharge and discharge zones.
- Contour groundwater levels, both elevation and depth below ground level, to determine boundary conditions, inflow and outflow points, the effect of surface water bodies on the system and a piezometric groundwater level map.

- It is essential that elevations of boreholes should be surveyed to at least 0.10 m accuracy for better interpretation of the water levels in relation to one another.
- Simulate groundwater level fluctuations with the Moving Average method and the Cumulative Rainfall Departure method depending on the type of available data.
- Risk of sinkhole formation should be quantified and management options should be described.

Groundwater quality measurements are used to understand the groundwater character in the dolomitic compartment and the following analyses must be performed:

- Diagnostic diagrams that would typically consist of Piper, Expanded Durov and Stiff diagrams that use chemical element ratios to plot different chemical analyses in different fields, of which each has a different interpretation and meaning.
- Spatial analysis is used to view and explain the various water qualities over an area in relation to local geological and geomorphological features, and contamination sources.
- Trend analysis is the plotting of time series of one or more chemical constituents in relation to seasonal rainfall patterns. This gives one an indication of the seasonal variation of chemical constituents compared to rainfall, whether it is ambient trends or contaminated sites.
- All these analyses can be done relative to a specific water quality standards, thereby also determining the fitness for use of the water.
- Environmental isotopes can be used where data is available to provide a signature of the origin of the groundwater type, identify mixing of different types of groundwater, provide information on through flow velocities and directions and to provide data on residence time, therefore the relative age of the water (Ford and William, 1989). However, care should be taken when doing the interpretations, as various influences could have been introduced onto the system.
- Water quality impacts must be quantified and included in the groundwater management plan for a dolomitic compartment.
- Health threats in an area must be quantified and the users informed accordingly e.g. contour maps of nitrate values should be compiled and reviewed on a continual base.

Various aquifer parameters can be determined with available data, of which recharge is the most important parameter for groundwater management in dolomitic areas. The most useful parameters in groundwater management include:

- Transmissivity – optional depending on what its intended use is e.g. regional numerical modelling.
- Storativity – used in most simulations and therefore necessary for a water balance.
- Recharge – essential for groundwater management in any dolomitic area. The confidence level at which recharge should be determined will be a function of the importance of the compartment and the availability of data. Methods for recharge determinations include the Chloride method, Moving Average method, Cumulative Rainfall Departure method, Saturated Volume Fluctuation method, Equal Volume method and environmental isotopes.
- Water users must be defined and the legality of the uses determined.
- Groundwater level and spring flow simulations must be performed before the groundwater management tool is developed.

- The monitoring network in a dolomitic compartment must be established, including groundwater level and quality, rainfall, spring flow and extraction monitoring.

Various maps can be produced and the detail on these maps should take into account the variation in expertise of its users. Essential maps for an area include:

- A geological map depicting all relevant geological features and boundaries, therefore also depicting the extent of the dolomitic compartment.
- A hydrogeological map with the most important geotechnical controls and monitoring stations depicted on the map.
- A topocadastral map with all relevant hydrogeological and institutional boundaries.
- A map in the management tool, with the recharge zones and protection zones delineated, in order that groundwater licenses can be linked to the map.

3.4 *Management tool*

The basic principle of a first order tool is to include the essential mechanism in an understandable format, which can be used by any groundwater manager. Developing a groundwater management tool is dependent upon which geotechnical controls are essential to the management of the dolomitic compartment and which controls are only beneficial. Therefore the tool cannot be developed before the hydrogeological evaluation has been completed and all essential controls have been defined and determined. Beneficial controls do not have to be included, but if the data is available and if it will be collected in future, it should be included, in order that the tool becomes even more useful. Inputs into the tool must be straightforward and outputs easily used and controlled by mechanisms that cannot be changed by a layman. However, should new information come to light, it should just as easily be modified by professionals, to include refined techniques, parameters or simulations.

The Groundwater Management Tool for the Schoonspruit dolomitic compartment was developed in an MS Excel environment. Figure 8 shows the Title sheet, figure 9 the Menu

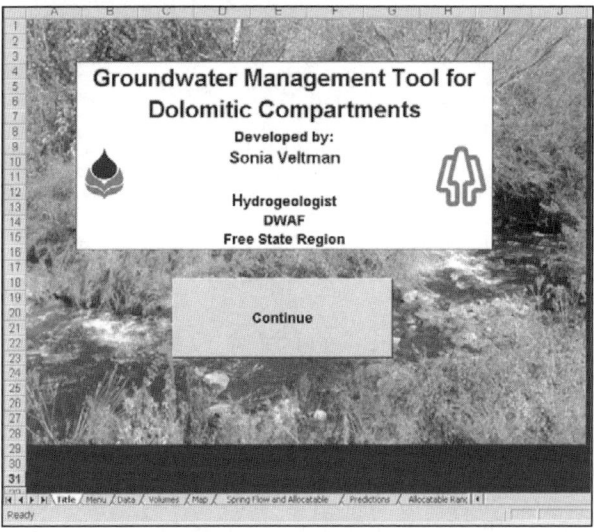

Figure 8. Title sheet of the Schoonspruit Groundwater Management Tool.

178 Groundwater flow understanding from local to regional scale

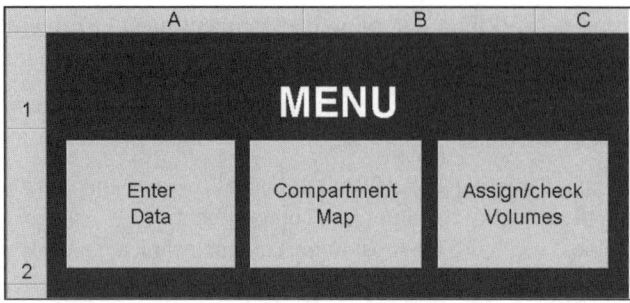

Figure 9.   Menu sheet of the Schoonspruit Groundwater Management Tool.

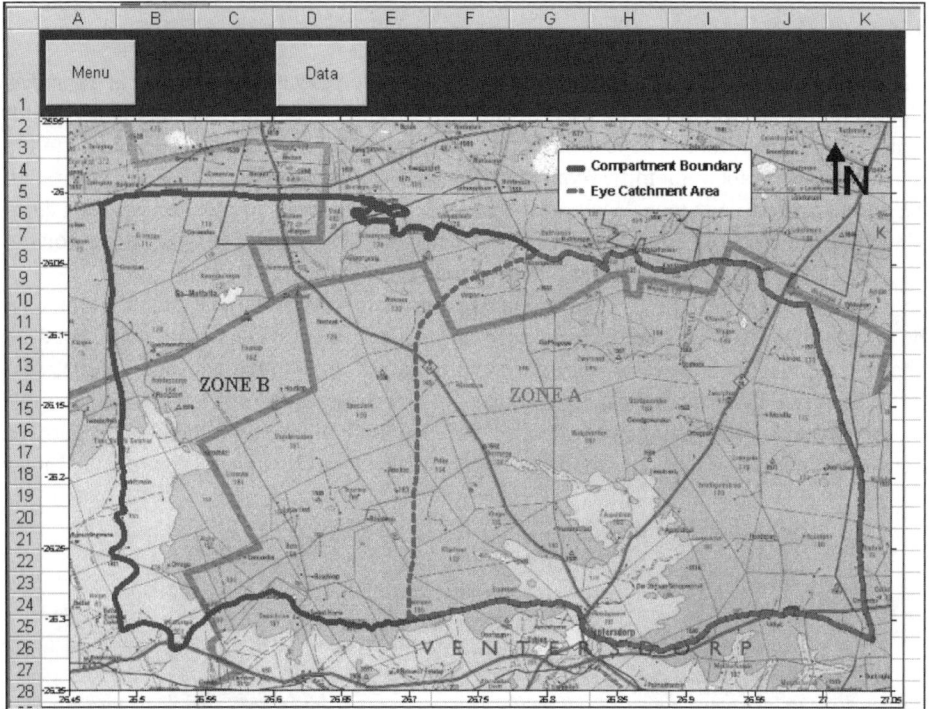

Figure 10.   Map sheet of the Schoonspruit Groundwater Management Tool.

sheet, with navigation buttons included, and figure 10 the Map sheet with two management zones delineated on the map. Essential geotechnical controls for the management of the dolomitic compartment was used to define the input parameters, figures 11 and 12, and the outputs from the analytical simulations in the tool are shown in figure 13 and 14.

Data input (groundwater quantities and qualities) is done in the yellow and orange cells, figure 11, yellow being compulsory data inputs and orange being optional data inputs. The optional data is helpful, if it is available, but the simulations are not dependent on these cells to run. Drinking Water Quality Classes for the different parameters have been added on the

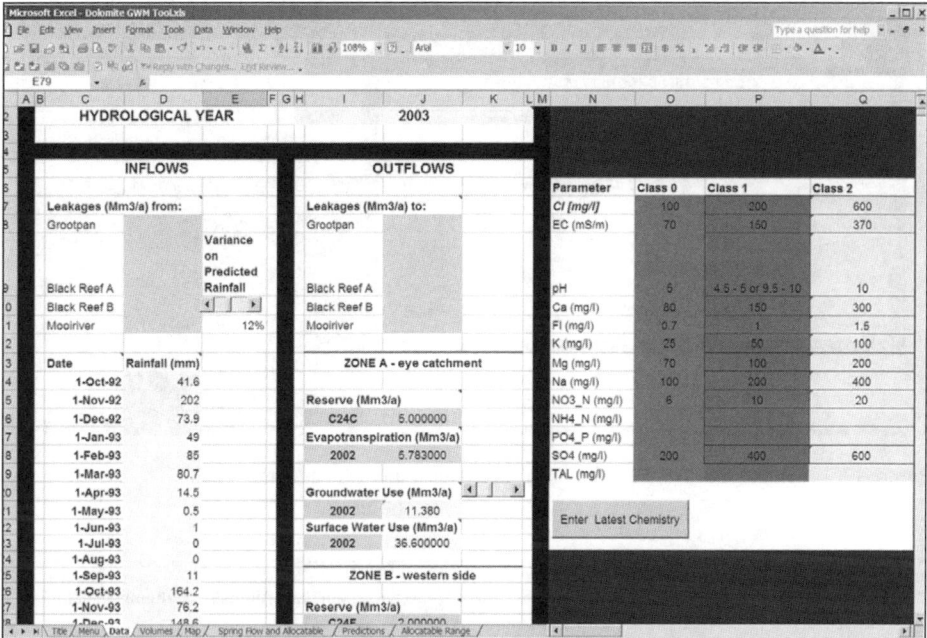

Figure 11. Input sheet of the Schoonspruit Groundwater Management Tool.

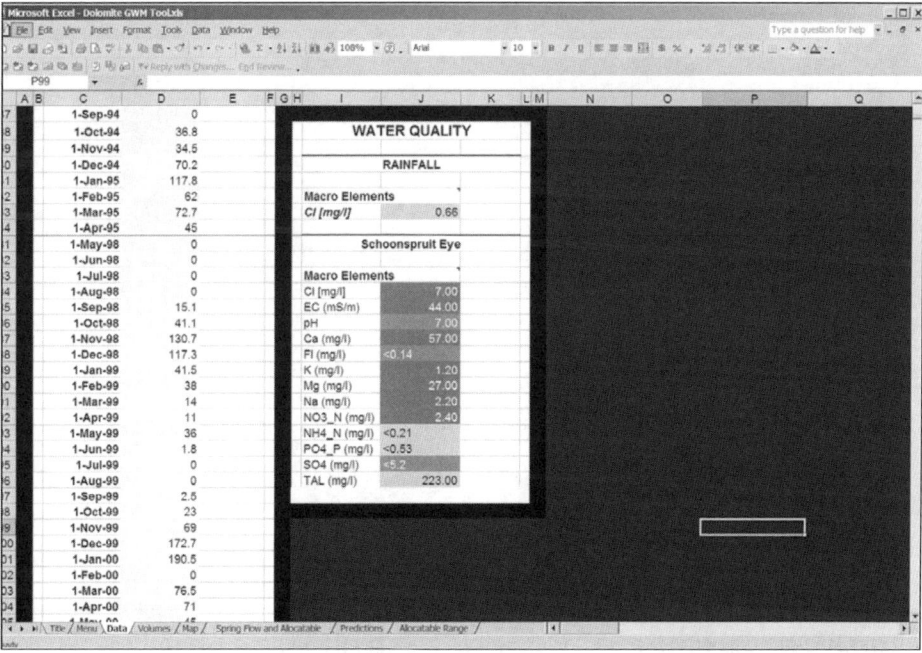

Figure 12. Chemical analyses of the Schoonspruit Eye.

180 *Groundwater flow understanding from local to regional scale*

	A	B	C	D	E	F	G	H I J K	L	M	N	O	P
1		Menu		Data									
2			RECHARGE TO THE SCHOONSPRUIT DOLOMITE							GROUNDWATER BALANCE			
3													
4			ZONE A - eye catchment			ZONE B - western side			ZONE A - eye catchment		ZONE B - western side		
5										Mm3/a		Mm3/a	
6		Type	Re %	Re Mm3/a	Type	Re %	Re Mm3/a			-2.460318		6.057632	
7													
8		Qualified Guess	7.00	38.741850	Qualified Guess	7.00	34.360331						
9		CI Method	29.33	162.346800	CI Method	9.43	46.281263						
10		H2 Isotopes	10.00	55.345500	H2 Isotopes	10.00	49.086188						
11		EARTH Model	6.00	33.207300	EARTH Model	6.00	29.451713						
12		SVF Method	5.50	30.440025	SVF Method	5.50	26.997403						
13		EV Method	4.90	27.119295	EV Method	4.90	24.052232						
14		Springflow MA	8.77	48.535330	CRD_MA Method	7.62	37.403675						
15													
16		Weighted Ave Zone A	8.77	48.535330	Weighted Ave Zone B	7.094714	34.825248						
17													
18-20													
21		Area Zone A (km2)		640									
22		Area Zone B (km2)		745									
23		Rain 96 month ave		54.90625									

Figure 13.   Volume sheet of the Schoonspruit Groundwater Management Tool.

	A	B	C	D	E
1	Date	Rainfall (mm)	Springflow (Mm3/m)	GW Abstractions (Mm3/m)	Allocable Volumes (Mm3/m)
2	01-Oct-92	41.6	2.28	0.6	
3	01-Nov-92	202	2.484	1.2	
4	01-Dec-92	73.9	2.26	1.2	
5	01-Jan-93	49	2.17	1.2	
6	01-Feb-93	85	1.93	1.2	
7	01-Mar-93	80.7	2.13	1.2	
8	01-Apr-93	14.5	2.05	1.2	
9	01-May-93	0.5	2.09	1.2	
10	01-Jun-93	1	2.02	1.2	
11	01-Jul-93	0	2.09	1.2	
12	01-Aug-93	0	2.09	1.2	
13	01-Sep-93	11	1.98	1.2	
14	01-Oct-93	164.2	2.131	1.2	
15	01-Nov-93	76.2	1.99	1.8	
16	01-Dec-93	148.6	1.96	1.8	
17	01-Jan-94	96	1.97	2.2	
18	01-Feb-94	123.9	1.87	2.2	
19	01-Mar-94	42.6	1.98	2.2	
20	01-Apr-94	39.4	1.87	2.5	
21	01-May-94	0	1.95	2.5	
22	01-Jun-94	0	1.9	2.5	
23	01-Jul-94	2	1.97	2.5	
24	01-Aug-94	2	1.97	2.2	
25	01-Sep-94	0	1.88	2.2	
26	01-Oct-94	36.8	1.91	2.2	
27	01-Nov-94	34.5	1.753	2.181	
28	01-Dec-94	70.2	1.69	2.181	
29	01-Jan-95	117.8	1.64	2.181	
30	01-Feb-95	62	1.48	2.181	
31	01-Mar-95	72.7	1.57	2.181	
32	01-Apr-95	45	1.57	2.181	
33	01-May-95	29.1	1.67	2.181	
34	01-Jun-95	0	1.61	2.4	

Figure 14.   Allocable Volumes of the Schoonspruit Groundwater Management Tool.

input sheet and these define the classes of the groundwater quality inputs in figure 12. Data inputs into the Data sheet include rainfall data for the 10 years prior to the current hydrological year, as well as predicted rainfall; various yields (Mm^3/a) of inflow and outflow, the storage, evapotranspiration, groundwater use and surface water requirements, and groundwater quality data for rainfall and groundwater in the compartment and in the Schoonspruit Eye.

The Volume sheet, figure 13, is a groundwater balance sheet dependent on the data inputs from the Data sheet, where recharge estimates from the different methods, including the spring flow simulation, have been added and weighed and an average recharge value can be obtained for each specific zone. The groundwater balance takes into account the volume of recharge, inflow and outflow, extraction and surface water needs.

Allocable Volumes, figure 14, are determined with the simple equation of subtracting surface water demand from the simulated flow, as this has already taken into account current groundwater use.

It is important to note that management of dolomitic aquifers does not necessitate the use of detailed modelling, but in this case it was much more useful to have a simple water balance approach to incorporate groundwater management into lower level management institutions.

4 CONCLUSIONS

Water managers and planners in other dolomitic areas need to follow the steps, as outlined in the *Technical Methodology*, in order to achieve a situation where the necessary information and data is sufficient to successfully implement groundwater management and planning in the respective areas.

A *Hydrogeological Evaluation* of the area must be done with current information and data before any new projects are initiated, thereby eliminating the possibility of duplication of work, as well as attaining a thorough understanding of how the system functions. If information or data is lacking, initiate the proper studies to solve the issues. When the aquifer characteristics and parameters have been defined, as outlined in the methodology, one can assimilate the necessary information and formulas into a workable groundwater management tool. In future interpretation of available data and information will lead to an improved understanding of local and regional flow systems and therefore the management of the dolomitic terrains in South Africa.

The *Schoonspruit Groundwater Management Tool* was an important deliverable of this research, since groundwater managers on the Schoonspruit compartment need to be capacitated to perform the most basic groundwater balances and simulations. Allocable volumes can now be determined on a continuous base for the 2 zones. The Drinking Water Quality Classes were introduced, as part of an early warning system, where drinking water quality is of concern.

REFERENCES

Bredenkamp, D.B., Botha, L.J., Van Tonder, G.J. and Van Rensburg, H.J (1995) Manual on quantitative estimation of groundwater recharge and aquifer storativity. Report no. TT 73/95. The Water Research Commission, Pretoria.

Ford, D.C. and Williams, P.W (1989) Karst Geomorphology and Hydrology. Unwin Hyman, London.

Kotze, J.C (1994) Summary of the Geology, Geohydrology, and Boundaries of the proposed SGWCA, District Ventersdorp, Drainage Area C24. Technical report no. 3833. The Department of Water Affairs and Forestry, Directorate Hydrology, Pretoria.

LeGrand, H.E. and Rosen, L (2000) Systematic Makings of Early Stage Hydrogeological Conceptual Models. GROUND WATER Vol. 38, No. 6, 887–893.

Mackay, H (1999) Water Resources Protection Policy Implementation: Resource Directed Measures for Protection of Water Resources, Integrated Manual, Version 1. Report no. N/28/99. The Department of Water Affairs and Forestry.

Parsons, R., Jolly, J., Titus, R. and Toksvad, T (2001) Strategies for inclusion of groundwater in the National Water Resource Strategy. Report no. 081 carl-2. The Department of Water Affairs and Forestry, Directorate Hydrology, Pretoria.

Polivka, J (1987) Geohydrological investigation of the Schoonspruit compartment in the dolomitic area of Ventersdorp. Technical report no. GH 3524. The Department of Water Affairs, Directorate Hydrology, Pretoria.

Author index

Adler, R. 151

Barthel, Roland 47
Bender, S. 73
Braun, Juergen 47

Cardona, A. 25, 85
Carrillo-Rivera, J.J. 25, 85
Cartwright, Ian 105
Castro-Larragoitia, G.J. 85

Dennis, Ingrid 135

Edmunds, Mike W. 1

Graniel-Castro, E.H. 85

Lawrence, C.R. 151

Mauser, Wolfram 47
Mieseler, T. 73

Ortega-Guerrero, M.A. 9

Rubbert, T. 73

Tweed, Sarah O. 105

Usher, B.H. 105

Veltman, S. 165
Veltman, Sonia 135

Weaver, Tamie R. 105
Wohnlich, S. 73

Subject index

Advection   2, 22
Aquifers
  Age
    Cretaceous   6, 32–33, 88
    Tertiary   32, 34–35, 39, 41, 57, 63
    Quaternary   11, 14, 32–33, 57, 59, 63, 79, 81, 107
    Jurassic   57–58, 63
  Alluvial   9, 14, 57, 59, 61, 88
  Boundaries   22, 25, 44, 148, 167, 169, 173, 175
  Coastal   1, 5–6, 87
  Confined   5, 36–37, 61, 118, 136, 142
  Contamination   42, 57, 75–77, 83, 85–87, 89–90, 96–101, 106, 124, 131, 135, 142, 146, 148, 175–176
    See also contamination and contaminant
  Dolomitic   165–166, 168–171, 174–175, 181
  Fractured   26–27, 32, 39, 60, 78–80, 89, 142, 147
    See also flow in fractures
  Geometry (extension, thickness, ..)   27, 33, 59–60, 68
  Granular   15, 21, 28, 34–35, 39, 59–60, 74, 78, 80, 89, 109, 142
  Igneous   1, 9, 11, 20
  Karstic   32, 50, 57–58, 60, 63, 165, 170, 178,
  Leaky (semi confined)   9, 28, 33, 38, 106, 142, 147, 151
  Limestone   32, 34, 58, 88, 106–107, 121, 123, 125, 127, 130, 154
  Management. See groundwater management
  Modeling   47–72
  Parameters. See Properties
  Perched   61, 88
  Properties   16, 31, 54, 88, 128, 141–142, 166, 168, 170, 173, 175–176, 181
  Porous. See granular
  Semi confined, See leaky
  System   2, 5, 60, 86
  Tests   16, 19, 27, 88, 109, 128, 142

  Unconfined   86, 105–106, 142, 151
  Volcanic   32, 34, 89, 153–154
Aquitards   9, 11, 13–17, 19–24, 28, 32–33, 36–37, 45, 106–107, 154
  Fractures in aquitards   20, 22–23
Attenuation   98

Bacteria. See Contaminants

Carbonates   127
Climatic change   1, 62
Climatic conditions   2, 64, 66, 125, 175
Coastal aquifers. See Aquifer, coastal
Connate waters   1
Contaminants   6, 22, 39, 142–143
  Bacteria   98, 100, 143
  Fertilizers   98
  Fluoride   35, 39
  Hardness   90
  Migration of contaminants   20
  Organic substances   100
  Pathogenic   98
  Viruses   143
  Wastewater   85–104
Contamination   57, 75, 85, 87, 90, 97, 98
  Diffuse   86, 96
  Point source   42
  Potential sources   77
  Risk of   76, 83

Denitrification   97–98
Dispersion   2, 98, 101, 142
Discharge (Zones, groundwater, aquifer, rates)   6, 11–12, 14, 22–23, 26, 31–36, 44, 50, 55, 61, 64, 78, 86, 88–89, 99, 106, 109, 118, 140, 148, 151, 160–162, 175
Dissolution
  Anhydrite   94
  Carbonate   120–121, 127
  Carbon   125
  Dolomite   119, 168
  Gypsum   105, 119, 130

Dissolution (*continued*)
  Halite   120–121, 125, 156
  Mineral   109, 119, 123
  Silicates   96, 125
Dissolved oxygen   36, 89–90, 92, 119
Drinking water   9–10, 22, 55, 57, 59, 62, 65–66, 86, 93, 97–98, 102, 142, 147, 178, 181

Environmental change   1, 48
Eh   36, 89, 98, 153–154

Flow
  Flow system   9–10, 12, 22, 25–46, 55–56, 60
  Fracture flow   6, 9, 20, 22–23, 26, 60, 78, 88
  Fractures   6, 9, 26–27, 31–33, 35, 39, 60, 80, 88, 141–142
    Hydraulic evidence of...   9, 17
    Influence on hydraulic conductivity   20
    Influence on contaminant transport   22

Groundwater
  Chemistry   3, 6, 21, 23, 25, 31–32, 36, 39, 85, 90, 94, 105, 109, 118–119, 121, 123–125, 127, 130–131, 153, 156
  Dating   3–4
  Development   1
  Flow pathways   1, 36, 86
  Isotopes   3, 5, 11, 21, 28, 105, 156, 170, 176
  Management   5–6, 9, 27, 45, 47–84, 135–149, 151, 162, 165–182
  Flow Systems   5, 9–10, 12, 22, 25, 27–28, 31–32, 85, 88–90, 102, 135–136, 142, 148, 157, 161, 165, 175
  Quality   1, 25, 85–87, 89, 95, 101–102, 144, 168, 173–174, 176, 181
  Residence time   3, 6
  Recharge, See Recharge
  Discharge, See Discharge
  Salinity   93, 105, 118, 122, 125, 154, 157
  Temperature   3, 30, 35, 90

Holocene   1, 3, 5–6
Hydraulic conductivity   10, 17, 19–20, 23, 27–28, 31, 42, 68, 76, 88, 102, 118, 121, 127–128, 152, 156, 162
Hydrochemistry. See groundwater chemistry
Hydrogeology   6, 32, 51, 57, 67, 88, 105–106, 109, 147, 153–154, 164, 175

Inorganic carbon   119
Isotopes   3, 5, 11, 21, 28, 105, 156, 170, 176
  Deuterium   11, 21, 121
  Oxygen   11, 21, 121
  Radiocarbon   3, 6, 125
  Tritium   22

Mineral dissolution   109, 119, 123
Modeling   12–13, 16, 21–22, 25, 30–32, 44, 47–71, 130, 165, 173, 176, 181

Noble gases   2–3, 6
Nutrients   87, 90, 96, 101

Organic matter   14, 22, 90, 101
Organic compounds   22, 87, 96, 100
Oxidation   77, 96, 101

Palaeowaters   1, 2, 5–6, 123
pH   36, 89, 98, 153–154
Piezometers   11, 19, 21
Pleistocene   1–3, 5–6
Porosity   16, 28, 60, 68, 78

Radiocarbon. See Isotopes
Recharge   1, 3, 5–6, 10, 12, 22, 26, 31, 36, 51, 53, 55–56, 59–61, 64–68, 74, 88–90, 94, 97–98, 101, 105–106, 109, 116, 118–119, 121, 123–125, 127–128, 130–131, 141–142, 147, 151–152, 156, 158, 160–162, 168, 170–177, 181
Residence time. See groundwater residence time
Risk analysis   80, 144, 146

Salinity   6, 21–22, 28–30, 32, 36, 41, 93, 97, 101–102, 105, 109, 118–119, 121, 123, 125, 129–131, 151–163
Sedimentary basins   1, 3, 5

Tracers   1, 3, 83, 123
Tritium   22

Urban areas   9, 22, 86–87, 89, 147

Vadose zone   96, 98, 100–101
Viruses. See Contaminants

Water-rock interactions   3, 125
Wells   7, 15, 55, 57, 79, 89–90, 100, 102, 116

SERIES IAH-Selected Papers

Volume 1–4 Out of Print

5. Nitrates in Groundwater
   Edited by: Lidia Razowska-Jaworek and Andrzej Sadurski
   2005, ISBN Hb: 90-5809-664-5

6. Groundwater and Human Development
   Edited by: Emilia Bocanegra, Mario Hérnandez and Eduardo Usunoff
   2005, ISBN Hb: 0-415-36443-4

7. Groundwater Intensive Use
   Edited by: A. Sahuquillo, J. Capilla, L. Martínez-Cortina and X. Sánchez-Vila
   2005, ISBN Hb: 0-415-36444-2

8. Urban Groundwater – Meeting the Challenge
   Edited by: Ken F.W. Howard
   2007, ISBN Hb: 978-0-415-40745-8

9. Groundwater in Fractured Rocks
   Edited by: J. Krásný and John M. Sharp
   2007, ISBN Hb: 978-0-415-41442-5

10. Aquifer Systems Management: Darcy's Legacy in a World of Impending Water Shortage
    Edited by: Laurence Chery and Ghislaine de Marsily
    2007, ISBN Hb: 978-0-415-44355-5

11. Groundwater Vulnerability Assessment and Mapping
    Edited by: Andrzej J. Witkowski, Andrzej Kowalczyk and Jaroslav Vrba
    2007, ISBN Hb: 978-0-415-44561-0

12. Groundwater Flow Understanding – From Local to Regional Scale
    Edited by: J. Joel Carrillo R. and M. Adrian Ortega G.
    2008, ISBN Hb: 978-0-415-43678-6